Quentin Hiernaux

From Plant Behavior to Plant Intelligence?

Conference-debate
organized by the group Sciences en questions
at INRAE Grand Est-Nancy
on May 21, 2019
and INRAE Clermont-Auvergne-Rhône-Alpes
on May 24, 2019.

Éditions Quæ, RD 10, 78026 Versailles Cedex, France

The Sciences en questions collection welcomes texts dealing with philosophical, epistemological, anthropological, sociological or ethical questions relating to science and scientific activity.

Editors of the collection: Raphaël Larrère, Catherine Donnars.

The working group was set up at INRA –now INRAE– in 1994 on the initiative of the departments responsible for training and communication. Its objective is to promote critical reflection on research through contributions that shed light on philosophical, sociological and epistemological issues related to scientific activity in an accessible and attractive form.

Translated from the French with the collaboration of Jeanne Allard from the original edition: *Du comportement végétal à l'intelligence des plantes ?*, Versailles, éditions Quæ, 2020.

This book was funded by the INRAE.
Its digital versions are licensed under CC-by-NC-ND.
https://creativecommons.org/licenses/by-nc-nd/4.0/deed.fr

Éditions Quæ
RD 10, 78026 Versailles Cedex
France

© Éditions Quæ, 2023

ISBN print: 978-2-7592-3745-6

ISBN PDF: 978-2-7592-3746-3

ISBN ePub: 978-2-7592-3747-0

ISSN : 1269-8490

Acknowledgements

I am grateful to the Fonds national de la recherche scientifique (FNRS) of Belgium for its financial support during the postdoctoral research which led to this publication. On a more personal level, I thank the INRAE Sciences en questions group for its invitation, and, more specifically, Sophie Gerber for numerous conversations and for reading an earlier draft of this volume. I am likewise thankful to Raphaël Larrère for his comments, suggestions and questions on previous versions. I also thank Catherine Lenne and Bruno Moulia for our discussions and, more generally, the INRAE centers in Clermont-Ferrand, Bordeaux and Nancy for their warm welcome.

Table of contents

Preface

It has long been known that plants are regarded by the public as still life. While it is accepted that plants do germinate, grow and flower, it is not immediately obvious to us that they do much else. However we ourselves are animals and we impose specific animal requirements on everything else biological. Our requirement to detect movement, for example, is limited by the nature of our visual process. The retinal image lasts about a tenth of a second and numerous, vibratory movements of the eye (usually unperceived) are necessary to prevent retinal adaptation. If the movement is much slower than these visual limitations the description of still life is obvious–but from our perspective. Some bamboos grow a metre a day but at less than a mm/minute, it is still not obviously visible. The consequence is that whereas animal behaviour is easily seen and deductions made about both its instigation and likely consequences, plant behaviour has always had to rely on experimental circumstances with appropriate measuring devices to establish that plants do really behave. And even then the half below ground remains largely invisible. Only with the onset of time lapse can many plants now be easily seen to be doing something; to behave. And to a much wider public. While Jane Goodall could record chimpanzee behaviour with merely a pencil and notepad, only with special cameras or other complex experimental apparatus could plant behaviour in wild circumstances be recorded. Much of real plant behaviour in wild conditions still remains unreported.

Plants are among the only groups of organisms that use an external source of energy; the sun. The consequence is that they are the basis of all food chains and predation of one kind or another, threatening survival, was inevitable from the time some two billion years ago when plants first separated from their protozoan ancestors. The evolutionary solution has been the construction of a plant body composed of repetitive elements, leaves plus subtended buds above ground and branch roots below. Inevitable loss of some simply leads to replacement by others. Growth takes place in embryogenic meristems in shoot and root tips. Furthermore predation and disease were tackled by the acquired ability to synthesize what is termed natural pesticides; substances that often flavour our food but do not kill us, because we are so much larger than any insect. The movement of plants to land some 500–700 million years ago subjected plants to additional environmental hazards. These are specifically sensed too and result in selective changes of the phenotype, often called plasticity. These changes are adaptive, designed specifically to potentially help survival, to continue growth of a kind and as far as possible reproduce. Plants are more sensitive to a much greater number of environmental signals that require adaptive change than the common roaming animal. Plants know about their environment because they respond to it; they are cognitive. Individuals control their own behaviour as cognitive agents to counter the hazards they perceive. Virtually all plant tissues are plastic. Plasticity is used to construct a phenotype with improved chances of survival, to fight over space and resources and construct a dynamic niche underground.

Biological intelligence is quite simply adaptive behaviour, improving survival probabilities as Dobzhansky indicated some 70 years ago. Easy to see when a zebra runs away from a marauding lion or chooses to continue movement to find un-grazed food. Plants approach similar goals when they synthesize a chemical to kill off marauding insects or choose to search new soil by root proliferation when phosphate deficiency is sensed. Animals move, plants change structure and physiology; the goal is identical. For those that like simple analogies; there are two kinds of cars on European roads, those run by electricity and those using petrol. But the goal, transport of people or goods is the same despite the entirely different mechanisms.

However the choice of words to describe plant behaviour, intelligence, agency, cognition, consciousness (or better awareness) and incorrectly believed by some to require nervous systems, creates controversy. This book by a young Belgian philosopher of science deals with many of these issues. Intelligence, memory, learning, consciousness are discussed in the first part. The second part concentrates on biosemiotics, how meaning is created from the perceived signs and signals that plants experience and it creates a plant ethology. There is an ongoing debate among plant scientists that will continue until plant physiologists doff their white coats and decide to understand how plants do behave in the real world. A place of environmental uncertainty, extreme competition, battles over space and resources, disease, invasion, common death and real predation in the many ecosystems of the planet. This book should interest and educate any open-minded scientist who wants to understand better the current controversy and the increasing under-standing of how complex, plant behaviour actually is.

Professor Tony Trewavas
FRS.FRSE.
University of Edinburgh

From Plant Behavior to Plant Intelligence?

Introduction

Behavior is a key concept in numerous fields of study: psychology, ethology, but also in the biology of organisms. It does not cause much surprise that dolphins, chimpanzees or rats display rational behavior–after all, they are not so different from us. But what about the organisms we deem "simpler"? Or even brainless organisms like plants? Do they display behavior at all? Is their behavior comparable to the behavior of animal species, or even to human behavior? Do plants give meaning to their environment? Do their activities result from a cognitive process? These questions are the starting point of recent controversies–of which the "plant intelligence" debate has received the widest coverage in mainstream media. But looking beyond controversies, we will see that it is essential to investigate plant behavior. According to the theory of evolution, life forms are continuous. Hence when we study behavior in biology, we cannot a priori, arbitrarily exclude some life forms from our investigation. We must bring forward and test justifications and arguments which will identify analogies in the behavior of distinct species, but which will also point to behavioral differences between them. Recent scientific experiments contribute to this inquiry. In philosophy, the study and interpretation of plant behavior will lead us to rethink concepts such as memory and consciousness, but also to reflect on the nature of the mind. Such a task will require subtle arbitration and a detailed examination of classical oppositions rather than catchphrases. Inquiring into how we use the notion of plant behavior will reveal a strained divide between reckless anthropomorphism and confirmed scientific reductionism–philosophy will allow us to examine it properly. Yet, to think beyond the anthropomorphism-reductionism divide turns out to be complex. This is why this work acknowledges that anthropomorphism can sometimes be of use to draw the attention to neglected topics – but that, doing so, it may cause distortions, for instance when it grants plants human emotions and attitudes. By contrast, reductionism studies phenomena solely through its observable causes, thus minimizing the risk of anthropomorphism – but, doing so, it often avoids the epistemological, ethical and metaphysical problems that lie at the foundation of biology (Canguilhem, 2008; Myers, 2015).

Let us first investigate the nature of behavior. What do philosophers and biologists mean by that? And what are the specificities of plant behavior? How could we distinguish it from the activities of a stone or from those of an animal?

These questions lead us to look deeper into problems where science and philosophy go hand in hand. They also require us to examine the often-hidden historical context in which these problems arose. Indeed, at least since the development of modern botany, philosophers and naturalists have been concerned with the nature of movements in plants and with the possibility of sensibility, and even soul, in plants.

Such interrogations are actually the core of the recent controversies on communication, memory, learning, consciousness, cognition and mind[1] in plants. We must reassess these notions, starting from a critical, better-informed standpoint. The present study will then more specifically put forward an original biosemiotic view of plant behavior. Finally, what does the recent excitement about these issues tell us? And

---

[1] Cognition and mind are sometimes used like synonyms. The way we use them here roughly echoes the difference between "mind" as a term used to refer to some abstract or metaphysical thing more relevant to philosophy, and "cognition" as a term scientists generally favor to describe the mechanisms of information processing.

to what epistemological and ethical developments can they lead us?

General Considerations on Behavior

Behavior is often intuitively approached from a human perspective. Yet, one can distinguish three distinct levels of behavior: a psychic level (which, in our intellectual tradition, is a priori taken to be typically human, even though scientists now recognize it in vertebrates and some cephalopods), a biological level (which concerns physiology) and a physical level (which concerns stones and particles). The level of behavior most commonly used to understand plants as organisms is the biological level. This section aims to distinguish the behavior of a plant from the behavior of a molecule or a human being, with the controversies about the differences and analogies between plants and animals as a backdrop. A clarification may first be in order: despite its pedagogical usefulness, the tripartition of behavior sketched here remains open to discussion.

Let us first distinguish the behavior of living beings from the behavior of non-living things. A preliminary, very general definition circumscribes biological behavior (displayed by animals, plants and other living beings) as an active response of the organism:

> Here we use the term behavior to mean what a plant or animal does, in the course of an individual's lifetime, in response to some event or change in its environment (Silvertown and Gordon, 1989, p. 350).

How is the motion of a stone following a shock different from a similar motion performed by a living being? Stones and other physical entities can only undergo events — it cannot respond to them. The nature of living activity, *i.e.* a living thing's response, must be specified. All organisms, including plants and unicellular organisms, respond to their environment according to internal processes. Since they depend on such internal mechanisms, the response is slightly delayed, unlike a stone's reaction to a shock, which is immediate. Internal processes are thus causes of behavioral responses (Dretske, 1988, p. 26–27).

Like stones, organisms can also undergo events. But, when they are sensible to them, when they process internally the information they obtain from stimulation and when they react to it in a delayed and observable manner, they display behavior. Thus, for there to be behavior, a reaction cannot follow uniquely from the stimulation without any mediation. But such a theoretical distinction is sometimes hazy when it comes to practice. Take this example. If I cut myself on a shard of glass and start bleeding, cutting myself is a behavior (since I wanted to pick up the shard of glass). By contrast, bleeding is not a behavior (since the wound is incurred without activity on my part). On another note, my organism's reaction to the wound and the coagulation of the blood following the wound does display a behavior. Furthermore, the activity or passivity of a behavior depends in part on one's perspective (Dretske, 1988). Take the following account of a miscellaneous news item: "Betsy was run over by a bus." Phrased in this way, it describes how someone underwent an event, and thus does not describe behavior. But if the account goes like this: "Carelessly attempting to cross the road outside designated crosswalks, Betsy was run over by a bus.", the tragic incident becomes the observable consequence of a behavior. Examples like this one show that behavior is relative: it always depends on an observer's point of view, on the context, as well as on the causal chain one takes into account. This aspect of behavior is crucial when we want to understand all the problems and controversies surrounding plant behavior. For what counts as behavior – even biological behavior – depends on an interpretation made within a specific theoretical framework.

Following the work of behaviorists and Tinbergen (1963), the methodology for the study and interpretation of biological behavior has primarily favored causes. It explains behavior through four main causes: its mechanisms, its functions, its ontogenesis (*i.e.* its development) and its phylogenesis (*i.e.* the evolution of the species involved). Based on the work of Watson and Skinner, behaviorism defines itself as the objective study of behavior. It relies on the observation of stimulus-response mechanisms, and opposes psychic explanations involving mental states. These behavioral approaches have prevailed in ethology, especially where "simple" organisms are concerned. As a result, traditional experimental ethology has hardly studied the perceptions, significations or motivations involved in animal behavior from their own standpoint. A different approach, however, is warranted in the ethological study of vertebrates such as monkeys, dolphins or dogs. In order to account for behavior in these animals, it proved necessary to grasp perceptions, emotions or motivations at the psychic level. Yet this non-reductionist approach can just as well apply to organisms more distant from us. The philosopher and ethologist Jakob von Uexküll theorized it in his biosemiotics using ticks and urchins (Uexküll, 2010). Biosemiotics is the study of biological significations, *i.e.* of the way an organism interprets the information provided by its milieu. Such an approach does not seek to reduce the study of behavior solely to its observable, causal dimension–it is non-reductionist. Following Uexküll's approach and its phenomenological roots, any organism is taken as a subject together with its own world, the stimuli from which it interprets. Even though material, functional, developmental and phylogenetic causes may explain behavior, such explanation does not do away with its effects on the animal, *i.e.* its experience of its environment. Biosemiotics tries to understand how an organism perceives and interprets its environment from its own point of view (not the point of view of its neuroreceptors).

The traditional approach can thus be said to focus on the "how" of a behavior, while biosemiotics also focuses on the "why" proper to the internal experience of a behavior. Biosemiotics is compatible with the study of reflexes, but it cannot be reduced to it. By contrast, because it limits its explanation of behavior to a reflex model, a reductionist methodology like some behaviorism's will not take into account an organism's own experience and significations. This is why reductionist behaviorism has for a long time avoided the observation of complex animal behavior. Since it refuses to associate behaviors with the elaborate cognitive capacities which result from a reflection and decision process, reductionism denies itself the means to examine such hypotheses and to create apparatus to test them (Despret, 2009). Strict behaviorism no longer reigns in animal ethology. But does this mean we can support a biosemiotics of plant behavior based on Uexküll's work? We will turn to this question in the last chapter.

Theoretical context greatly influences the way we approach behavior. It is then no surprise that the history of ethology and botanical sciences turns out to be essential to a better understanding of the current status of plants and the controversies surrounding it—all the more so since such history is itself strongly influenced by our cultural and philosophical heritage concerning plants.

Botanists and Philosophers on the Activities of Plants: a Historical Approach

During the second half of the 20th century, several rounds of reports on experiments on plant sensibility have appeared in mainstream media, generating misinformation and sometimes nebulous theories. It was reported that plants were sensitive to music—even showing a preference for Mozart rather than hard rock—that they emitted healing waves in the nearby air, etc. According to one of these iconic theories, Cleve Backster (1966), a former CIA agent with no scientific education, performed experiments allegedly demonstrating telepathic abilities in plants. Plants had been shown to react to an experimenter's malevolent intention to boil a shrimp–a lie detector connected to the plant proved it!

These so-called experiments have been disproven by the scientific community and no one has been able to replicate their results. Nonetheless, they teach us several things. First, there is an enthusiastic public who wants to believe in the hidden powers of plants–as shown by the commercial success of books concerning these experiments. It is too simple to credit the fascination for New Age and pseudoscience for such excitement, because it also shows an attachment for plants and the will to know more about them. Second, the fear of being associated with notorious hoaxes has brought some discredit on the study of plant behavior. Biologists doing earnest and reliable work on plant sensibility were then tempted to censor themselves. As a result, the rare researchers who were interested in such issues have often based their experiments on a rigorous, but reductionist stance.

At the start of the 21st century, a novel view of plant behavior emerged in research labs. It was inspired by animal ethology and less reductionist. It called itself "plant neurobiology" (Brenner *et al.*, 2006; Barlow, 2008; Mancuso and Viola, 2015) and it shook the community of plant biologists. Strongly polarized, vivid debates developed around new interpretations of the mechanisms of chemical, hydraulic and electrical signaling in plants. But the research projects of plant neurobiologists also suggested a more fundamental turn: to reinterpret experimental results in terms of communication, memory and even intelligence. Such notions were previously exclusive to humans and so-called superior animals (Trewavas, 2003, 2004, 2005, 2014; Firn, 2004; Alpi *et al.*, 2007; Brenner *et al.*, 2007; Cvrcková *et al.*, 2009; Calvo and Baluška, 2015). As recent publications show, the debate is far from over (Taiz *et al.*, 2019; Calvo and Trewavas, 2020).

From a historical perspective, what are the scientific reasons for such a divide between animals and plants in the study of sensibility and behavior? If we investigate the arguments made during the course of the history of botany, would we not find them to be chiefly philosophical or ideological in nature? In any case, a plurality of interpretations seems to have existed throughout history (Hiernaux, 2019).[2]

The scientific study of plants begins in Antiquity with the philosopher Theophrastus, a disciple of Aristotle, now held to be the founder of botany. In the very first lines of his *Enquiry into Plants*, Theophrastus decidedly states that "conducts and activities we do not find in [plants] as we do in animals" (I, 1, 1). Despite an acute sense of observation and a knowledge of plants unmatched in his time, Theophrastus remains under the influence of his master. Aristotle is well known for having clearly distinguished plants from animals and humans based on the faculties of their soul. In Aristotle's *De Anima (On the soul)*, plants are only able to grow and get nourishment, while animals also possess sensibility, and humans, rationality. Key premises of the Western hierarchy of being and life (Lovejoy, 1964) go back to Aristotle. Thus, while Theophrastus had already noticed connections between species and their habitat, and observed movements in some plants, he also seems to accept that plants, differently from animals and humans, have no sensibility, emotion or desire[3] and are, consequently, deprived of any intellectual faculty.

Theophrastus's works are then lost in the West, and the life sciences do not make much progress during the Middle Ages. Only Aristotle's texts are circulated, and his treatises on animals are authoritative sources. Theophrastus's works on botany are rediscovered, translated and printed only in the Renaissance (Greene, 1909; Magnin-Gonze, 2009).

---

[2] This paper goes into greater technical details on the history covered in the following section.

[3] Plato was more inclined to acknowledge desires, pleasures and pains in plants.

While Aristotle identified a variety of movements (like growth, corruption or alteration), the modern period reduces movement to perceptible local transportation and animal locomotion (Marder, 2013a, p. 21). Modern philosophers and botanists thus point to the motionlessness and passivity of plants, deeming these features crucial for the hierarchical distinction of plants and animals. Plants are considered to be just barely living–and behavior is denied to them.

However, our senses–which do not perceive movement reactions in plants as they do in animals–agree only imperfectly with what is observable from a scientific perspective. As early as the 16th century, several botanists (Cordus, da Orta, Gesner) describe movement in the leaves of legumes and others. But how can one account for these activities if plants are held to be insensitive? Here, theoretical framework and observation seem to be inconsistent. Never mind that–modern naturalists attempt to explain movement in plants while staying true to the theoretical principle of their insensibility. Plants are thus said to drink water the way sponges soak it up, or to be drawn to the sun the way iron filings are drawn to a magnet–*i.e.* physical mechanisms would determine a plant's movements externally, and there is no need to posit activity and sensibility in the plant itself. But this hypothesis, while theoretically elegant, turns out to be inadequate when it comes to facts. The perplexity of the botanist and philosopher Cesalpino (1519–1603) on the subject testifies to this: he observes that neither the force of the void, nor the dryness of stems can account for a plant's intake of water (1583, p. 3–4).

We have a dilemma. Either we must admit that plants are sensible–and go against two thousand years worth of intellectual traditions. Or physical reductionism allows us to explain plant activity through determinist mechanistic laws–which remain unknown and mysterious. Since tradition and the *scala naturae* (chain of being) sometimes have more weight than a simpler explanation, the majority of modern botanists have favored the mechanistic option. Incidentally, the mechanism inspired by Descartes – which claims to make sense of the world solely through movement and matter – was not without its virtues. It led to progress in 17th-century botany through the work of experimenters like Mariotte, Hales or Ray (Delaporte, 2011). Even though numerous questions remain unanswered, some hydrostatic mechanisms involved in sap absorption and circulation substantiate the hypothesis of a plant-machine. This leads one of the most eminent of 18th-century botanists, Linné (1707–1778), to write: "Minerals grow. Plants grow and live. Animals grow, live and sense" (Linné, 1736, aphorism 3) in a thoroughly Aristotelian perspective. Linné's aphorism 133 confirms this insensibility of plant life: "Although plants are devoid of sensations, they live as much as animals do" (Linné, 1736). The philosopher and botanist Julien Offray de La Mettrie (1709–1751), a successor of Descartes, develops a more radical botanical mechanism. Based on the motionlessness of plants, he deduces a lack of sensibility in them – like his predecessors – but he goes further, and associates an organism's degree of intelligence with its degree of movement (La Mettrie, 1748, p. 43–44). Like other 18th-century naturalists, La Mettrie brought animals and humans together based on their instincts and their faculty of locomotion – and drastically separated them from motionless, insensible and thus stupid plants. The lack of instinct in plants was not only due to their lack of movement, but also to the lack of sexuality and mating in them. Even when the idea of plant sexual reproduction gradually became accepted during the 18th century, it was still often taken to be a strictly mechanical process, entirely different from the voluntary reproductive activity of animals.

The association of movement with intelligence is not a theoretical necessity grounded in philosophy or science. Yet it created an argument from authority which has lasted until our times. It appears, for instance, when the cognitive science philosopher Patricia Churchland (cited in Calvo, 2016) says that a rooted plant can afford to be stupid and take life as it comes (Churchland, 1986, p. 13) unlike animals whose movements provide for their bodily needs (Churchland, 2002, p. 70).

Philosophical and moral literature has shown a profuse interest in animals, as well as in the specific difference of humans, which it has relentlessly questioned and put into perspective (Afeissa and Jeangène Vilmer, 2010). But it does seem as though the true ontological divide of our tradition, which has remained mostly unexamined, is the one between animals and plants. After all, even if we are humans, we are above all animals. We have numerous features in common with the latter, obviously. But plants? There is a world of difference between us and them – or, more precisely, a kingdom. Modern philosophers and botanists realized that to compromise this divide would call into question a whole metaphysical and moral structure. As François Delaporte (2011) explains in his philosophical history of modern botany, to grant sensibility to plants does not only require us to undermine scientific and philosophical orthodoxy. It also requires us to reflect on the possible suffering of plants, and therefore on the way humans justify treating plants as they do.

In the 18th century, a very peculiar plant is brought to Europe. It starts to excite the interest of botanists such as Duhamel du Monceau, and Desfontaines. This plant, the *Mimosa pudica*, commonly known as the sensitive plant, displays extraordinary faculties. Whenever it is touched or undergoes any shock, it quickly folds up its leaflets and lowers its petioles alongside its stem. Desfontaines even apparently observes that, when it is moved about in a vehicle, the sensitive plant adjusts to the movement and reopens its leaflets despite enduring shocks from the continuous bumps in the road. And anesthetic substances inhibit its reactions. But even if these observations made the case for the sensibility and behavior of plants, they were generally taken to be exceptions proper to this species.

During the 19th century, the situation gradually progresses: plant physiology enters academic institutions and the sensibility of plants can be investigated on new terms. A.-P. de Candolle (1778–1841) is among the botanists who take part in the evolving debate. When he conducts experiments on the sensibility of plants to light, Candolle gives scientific bases to the idea that all plants may possess sensibility. But he goes farther: when he studies plant circadian rhythms,[4] which are expressed by regular periodical movements, he also points out the possibility of changing these rhythms. If we interchange daylight and nighttime, along with the origin and duration of illumination, the rhythms and movements of plants change accordingly. Moreover, once the plants have become familiar with these new rhythms, they persist for several days after original conditions are restored, testifying to a kind of "memory" in plants.

At the time, the main problem is that there are no known mechanisms to account for such behaviors. There are no nerves or muscles which could explain sensibility or irritability in plants, much less a brain which could house a memory. For many thinkers—including important botanists like Lamarck—the lack of mechanisms is the basis for an argument which warrants the rejection of plant sensibility. But, as some naturalists noted, the reasoning is not as self-evident as it first appears:

> For is it not obvious that it is not the absence or presence of nerves which shall decide whether a being possesses or possesses not the faculty of sensing, but rather some of the striking events of its existence, or the whole set of manifestations which make up its life? […] one could rightfully suppose so, even if nerves were absolutely lacking. (Boscowitz, 1867, p. 159–160)

---

[4] Circadian rhythms are biological cycles lasting approximately 24 hours. They manifest diversely in all organisms.

In short, one must not mix up the existence and purpose of sensibility with the means of its realization in animals. This leads advocates of plant sensibility to offer two clashing hypotheses. Either plants possess structures which are similar to nerves or work as they do – like "spiral fibers", for instance. Or there are mechanisms entirely different from those known to exist in animals. The discovery of such mechanisms, however, was difficult, since scientists did not have any reason to search for something which they thought should not exist.

Until the 20th century, botany stands out first and foremost as a descriptive science – even though plant physiology thrives as early as the 19th century in pioneering centers like Germany, and gradually wins acclaim throughout Europe. It is in these circumstances that the two hypotheses on the mechanism of plant sensibility develop.

In fact, there are no structures truly analogous to nerves in plants. As early as the 1880s, Charles Darwin and his son Francis (Darwin and Darwin, 1880) conducted experiments on phototropism which suggested that a hypothetical chemical substance must be the source of the bending movement in the stems (Bernier, 2013). Auxin and several other substances affecting plant activity were discovered afterward. Chemical messages, however, do not account for all kinds of sensibility and motion. As early as the 19th century, Burdon-Sanderson and Bose brought to light the existence of electrical currents in plants by studying the sensitive plant and Venus flytrap (*Dionaea muscipula*) – a carnivorous plant which triggers a swift trap to catch insects (Chamovitz, 2013, chap. 3). Some elements at work in plants could then serve a similar function as nerves in animals. Darwin himself likened the movement of growing root tips to a kind of decentralized animal brain performing effective and purposeful operations. The analogy mostly stops there, however. The successful results of experiments studying plant hormones urged biologists to pursue the chemical path rather than the electrical one in their investigations. But the chemical discoveries from the end of the 19th century did lead to a re-examination of the idea that plants were, in the words of the plant physiologist Julius von Sachs, "mere product[s] of the activity of matter, and so an unsubstantial appearance in the general circulation of nature, the offspring of blind agencies" – for physiology, even in plants, cannot be reduced to physics, chemistry and the "blind necessity" of mechanism (Sachs, 1890, p. 180). While he does reject vitalism,[5] Sachs points out: "Further, the purpose of the movement of nature must remain an insoluble enigma in this scheme of blind necessity" (*ibid.*) This concession suggests that the understanding of behavior cannot be reduced, even in plants, to the sole study of its causes and mechanisms.

In the 20th century, progress in the physical sciences arguably plays a part in the birth of behaviorism. It brings about a new, reductionist framework for the study of behavior, using reflexes, nerves, and thus animals as models. Scientists and philosophers discredit earlier work on plant sensibility. They bring back the clincher: that there should be no authentic behavior without a nervous system. Animals move, act and react to their environment – but this is merely the result of reflexes, hormones or genetic programming, they say. Following Morgan's canon (1894), there is no need to introduce autonomy, a mind or even cognition to account for the activities of an animal, if we can interpret these activities as belonging to lower faculties. Since we are bound by scientific rigor, the most parsimonious hypothesis – reflex action – must be preferred to others when accounting for behavior (Struik *et al.*, 2008).

But the most parsimonious hypothesis is not necessarily the most fitting. While some of the first behaviorists to theorize animal behavior did not make such a leap, the conflation of

---

[5]   Vitalism postulates that a metaphysical principle overlooks the organization of living beings.

the way in which we account for phenomena with reality spread like wildfire. From the original idea that no mind was necessarily required to account for animal behavior, we jumped to the idea that animals necessarily had no mind. But, as Daniel Dennett (1983) noted, reductionism may not be the most judicious method in biology – especially where behavior is concerned.

The law of parsimony calls for us to choose the simplest explanation. But it must not be mixed up with reductionism. The problem of behaviorist ethology is to have wrongly thought that parsimony required us to reduce behavior to mechanistic schemas (like reflexes). But such schemas some-times turn out to be less parsimonious than hypotheses based on processes or non-reducible phenomena like proprioception in plants,[6] and emotion or intention in animals, both to explain and to describe behavior.

The theoretical framework of behaviorist ethology therefore helped construe living beings as machines. It also helped widen the ontological divide between animals and humans, and between animals and plants. The discovery of DNA structure in the 1950s consolidated this framework with genetic determinism. Behavior, especially in the simplest organisms, was said to be programmed by genes (Jacob, 1981). More-over, advances in molecular biology led scientists to focus their investigations on the cellular level (and bacteria) at the expense of the (multicellular vegetal) organic level – at which behavior is most clearly displayed (Fox Keller, 1999, p. 254).

---

[6] Proprioception is a kind of internal knowledge of one's own body (cf. the section on consciousness). The discovery of plant proprioception in the 21st century resulted from investigations led under parsimony imperative. Numerous hypotheses had been formulated and numerous experiments had been conducted before coming to the idea that proprioception was the simplest hypothesis. Parsi-mony thus does not preclude the investigation into more elaborate capacities when a behavior is too complex, and when simpler mechanisms cannot explain it.

Correlation admittedly exists between some behaviors and some genes, notably the one found in insects like bees (Rothenbuhler, 1964). But during the second half of the 20th century, biologists realized that genetically programmed behavior is rare, more complex and determined less unequivocally than what was once thought (Atlan, 1999; Lapidge *et al.*, 2002). In fact, the determinist idea that plant behavior is thoroughly programmed turns out to be highly problematic, since behavior manifests mostly through growth in plants. But the growth of a plant, unlike the growth of an animal, is very much undetermined. It depends on the meristem, a kind of tissue mainly found at the top of stems and at the tips of growing roots,[7] where cells continuously differentiate and multiply. Plant organogenesis also depends on contextual factors. Meristems are not independent machines. For instance, the meristematic cells of the root cortex[8] differentiate according to information provided by the surroundings, more differentiated cells about their location–and not according to "intrinsic [...] information" (Raven *et al.*, 2013, p. 574).

In the 21st century, it does seem as if plants' unified, consistent behavior cannot be entirely reduced to genetic determinism, nor to the mechanism of each of its parts working like the cogs of a clock, or the gears of a machine.

Behavior in Plants

Given this historical contextualization, how should we interpret recent scientific experiments on plant behavior? What epistemological and philosophical problems do such interpretations pose? And how should we solve them?

---

[7] These primary meristems are involved in growth by elongation. In woody plants, there are also secondary meristems, which are responsible for the growth of stems and roots in diameter.

[8] The cortex (or cortical parenchyma) is a tissue found at the center of stems and roots. It comprises the meristematic cells from which it comes.

To this day, the theoretical framework for plant behavior is still most often limited to the study of causes and physiological mechanisms. But can we posit and study plant behavior as exclusively biological – as involving no "psyche" or cognitive skills whatsoever? In recent years, an explanatory pluralism favoring a kind of animal psychism developed in response to mechanistic behaviorism (Despret, 2002, 2009, 2014; Burgat, 2006, 2010). A situated inquiry into the related phenomena and controversies should tell us whether it is possible and legitimate to extend this interpretation to plants. But if we uncritically transpose terminology, we run the risk of merely animalizing or anthropomorphizing plant behavior. How can we understand the specificity of plant behavior? To what extent can we draw on animal ethology and biosemiotics to interpret plant behavior? And what problems does this pose?

Dissociating Being and Action

Our first obstacle is the following. In plants, there is no clear distinction marking out behavior among the whole of biological, physiological or bodily processes:

> Among plants, form may be held to include something corresponding to behaviour in the zoological field. (Arber, 1950, p. 2–3)

> In fact, growth in plants serves many of the functions that we group under the term "behavior" in animals. (Raven *et al.*, 2013, p. 539).

This gets us into a first theoretical difficulty because our intellectual tradition and its epistemological framework tend to separate body and mind (or soul). Even though the sciences now acknowledge that thought is correlated with brain structure, we implicitly persist in dissociating being and body from action and mind. We assume that behavior only ever results from the action of the mind on the body. The activities we deem strictly corporeal or biological, like reflexes or growth, would not count as behaviors, because they do not depend on the mind's action. However, the claim that plant behavior manifests in its form and growth questions such traditional dualism.

Individuality of Behavior

The second difficulty has to do with the scale and temporality we rely on to apprehend a unit of behavior. For instance, an animal reflex can well appear localized in its mechanism, but it has meaning only in relation to how it serves the individual as a whole (Canguilhem, 1965). A unit of behavior also matches a unit of time, *i.e.* the lifespan of the organism. From an evolutionary standpoint, the most relevant unit of behavior can be reduced to the unit of selection, *i.e.* the entity which survives and reproduces, thus maximizing its progeny through heritable variations. For instance, in numerous animals, the unit of selection is the entire organism insofar as its parts cannot reproduce on their own.[9] In the case of plants, however, it is difficult to delineate the unity or individuality of an organism, because plants are not centralized like animals. Besides the fact that intuitive limits are not always clearly perceptible in plants, their functional integration does not depend on a central nervous system and their organs are autonomous in a way in which an animal's organs are not. Thanks to totipotent cellular tissue, plants can regenerate, and even recreate a whole other plant from a part of their body. Is a shoot sprouting from the roots of a parent plant a unit of behavior if it remains connected to its parent? Since the plants concerned form a genetically homogeneous clonal population maintaining physiological relations – like super-organisms such as colonies of social insects – the scale most relevant to understand such behavior would not be the one of the individual plant, but rather the collective scale, according to some authors. The reverse may also be true: since different

---

[9] In some colonial invertebrates like coral or jellyfish, that is not necessarily the case.

elements of a plant – part, organ, or even cell – can produce a whole new plant which can inherit a new genetic mutation from these elements, plant cells or meristems may be the unit of selection, and thus the privileged scale at which we should study plant behavior, at least from an evolutionary standpoint. This is only one example among the many biological models for the coordination of the whole with its parts, of the individual with the collective in plants (Clarke, 2010, 2012; Gerber, 2018). The lack of individuality in plant organisms accounts in part for the privilege given to an infraorganic or supraorganic physiological model over an ethological model.

Plant activities have often been reduced to heritable adaptations. They would not be true behavior because their cause is not found in the organism's experience, but is rather the genetic result of natural selection. The adaptations plants display in a given situation would only constitute passive actualizations of the potential of their species' gene pool. If we extend this explanatory stance to any activity, it opposes the very possibility of learning, which is precisely the active adaptation performed by an organism during its lifetime based on past experiences. Traditional adaptationist arguments thus take learning to be a complex faculty, typical of animals and humans.

Agency in Plant Behavior

This being said radical adaptationism does not succeed in denying all agency to plants.[10] For it does seem that their biological activities are not strictly genetically determined (neither are they physically determined, as some modern botanists thought), or at least not any more than animal or human behavior is.

---

[10] To grant agency means to accept that a being possesses an autonomy of its own, allowing it to undertake some actions. It also means that such being possesses some minimal experience of its own life.

Furthermore, we must avoid the following misunderstanding. The determinate genetic structure which makes a behavior possible over the course of evolution (for instance, whether a plant develops adventitious roots)[11] is not identical with the effective manifestation of such behavior, which is not necessarily determinate (for instance, whether a plant develops one, two or five adventitious roots, and whether it does so in one direction or in another). Some activities may be genetically programmed – like the fact that a plant will flower when it has received a given amount of light. But this does not mean all plant activities are programmed in this way. Some other activities thus display behavior at the level of the organism – like the number of flowers in a plant, or its growth, both of which have little determination. Incidentally, some biologists reject more and more openly the claim that a nervous system or movement is a necessary condition for behavior (Silvertown and Gordon, 1989; Cahill, 2015, Introduction). Some put forward a biology of plant behavior involving choices and decisions related to the environment (Hodge, 2009). The use of choice and decision implies that plants are not thoroughly determined in their actions, for there are alternatives open to them. Choosing one option over another is the result of a "rational" process insofar as some options benefit the plant more than others (Cvrcková *et al.*, 2016). Choice is not a mere indeterminate alternative – as it would be in the case of a random physical phenomenon, for instance. It implies that the organism making a choice has a minimal value scale,[12] *i.e.* it tends to persevere in its being by taking into account at the very least what matters for its survival and reproduction. We would do well to remember that several distinct explanations of behavior coexist in scientific literature.

---

[11]   "In botany, an adventitious root is a root or rootlet growing directly on the stem in an accidental, fortuitous and unusual manner." <https://www.aquaportail.com/definition-3731-adventive.html> (retrieved on January 27, 2020).

[12]   Value is what directs choice toward one alternative, and so is distinct from randomness.

Our modern tradition has long denied plants biological agency by accounting for them as passive elements of the physical world. Here, what is at stake epistemologically is our understanding and classification of behavior as a continuum of faculties going from the most external and pragmatic descriptions to abilities which call for more or less developed internal agency in an organism. Studying plant behavior requires us to introduce more detailed degrees in such faculties. For instance, the use of choice–minimally sketched above–does not require us to introduce intelligence or consciousness, but rather only a basic degree of agency.

Among living beings, there seem to be motivations or autonomous conducts which manifest at certain scales and at certain moments–but not at others. Thus, plants direct their conducts, and have needs and goals for their actions. Likewise, the choice of one alternative over another in a plant's behavior shows that these alternatives have an instrumental value for it–from the plant's point of view, an alternative is bad and the other one is good. But concepts like choice or value cannot be reduced to a causal mechanism. The point here is that some behavior–at least when it presupposes a choice–implies some signification (even if it may not be represented as such) or some value that can be granted to different alternatives involved in the action. Gibson's affordance theory[13] (1979) and biosemiotics tend toward this idea, to the extent that, in these views, its habitat never appears neutral to the organism experiencing it. On the contrary, an organism's milieu has features which can induce action preferences. Such preferences then depend on what the organism perceives, or even on what it anticipates, and on the value or meaning the organism gives to what it perceives or anticipates. In all living beings–even unicellular organisms–the organism must at least be able to use constant regulation in

---

[13]   Affordance refers to all the features of an object or milieu which an individual can use to perform an action.

order to preserve the membrane which separates it from the external environment. In this sense, any living being must have some notion of what matters for it and of what threatens it – in this would lie a proto-mind, some minimal form of interiority (Jonas, 1966).

Cognition and Representation

Today, most scholars in philosophy of mind and cognitive sciences consider that plants indeed have sensations. Plant sensibility is now confirmed as a biological reality (see Trewavas, 2014 for a synthesis). But they also consider that plants have no perception proper, and thus no mind. For to perceive one's environment would require ordering sensations through a mental act of representation. Since plants have no brain, they would be incapable of such feat (Maher, 2017). A new question then comes up: can we have some form of cognition or mind without mental representation? Most biologists and philosophers explicitly or implicitly answer no to this question, invoking cognition's ties with the central nervous system and brain. It is a valid theoretical stance to connect mind and mental representation. However, such an assumption does not exclude the reverse theoretical possibility – that cognition and mind can obtain without representation. Just as nerves were deemed to be the sole possible cause and the necessary condition of sensibility in the 19th century, representation may be deemed the sole cause and condition of cognition and mind. In this "standard" view, we cannot conceive cognition from a strictly externalist standpoint. Cognition indeed requires internalized mental representations, as well as the faculty in the observer to attribute such representations to some species, but not to others. At the very ground of this lies an epistemological debate on the role and scope of analogy in knowledge acquisition. For the strictest of behaviorists, the functional analogy with the human mind cannot be applied to any other living species.

We must, however, take note of the occurrence of such analogy within the human species itself: I can testify to the reality of my own mental states – but I can only testify to the reality of similar mental states in fellow humans because of my confidence in analogical reasoning. For, even though language allows humans to communicate more directly concerning their mental states, nothing stops me from thinking that the behavior of other humans results from complex programming simulating mental states similar to mine. Thus, if we a priori deny all analogical reasoning its legitimate place in the study of behavior, we are logically led to solipsism.

We must, however, distinguish an analogy from a simple metaphor. Where biology is concerned, an analogy compares two things which differ in nature based on some function, purpose or common structure,[14] which is real and can be verified via experimentation. An analogy does not imply that the things it compares are identical, precisely because it always concerns things which differ in nature. Metaphors are different: a metaphor is a linguistic comparison that does not necessarily imply a real community between its objects, and it can be based solely on the figment of one's imagination (as in a catachresis, for instance). A metaphor starts from words to describe things, while an analogy starts from facts to which it applies words. But the distinction is not as absolute a divide as it seems: what is first a simple metaphor can turn out to be an authentic analogy as our investigation progresses. This is why metaphors play an important role. They can indeed orient the search for authentic analogies – if we go beyond

---

[14]   If this structure comes from a common evolutionary heritage, we speak of homology rather than of analogy.

words.[15] Confusions between analogies and metaphors as well as debates on their epistemic roles in biology have therefore naturally entered discussions of the issue of plant behavior and cognition (Brenner *et al.* 2006, 2007; Alpi *et al.* 2007; Trewavas, 2007; Struik *et al.*, 2008).

Minimal Cognition

But we can also highlight the possibility of plant cognition based on the study of the effects of behavior. That is what is at stake in some of the contemporary research drawn toward minimal cognition (Van Duijn *et al.*, 2006; Calvo and Baluška, 2015; Godfrey-Smith, 2016). If we could point to manifestations of communication, memory, learning, consciousness and intelligence as we observe them in organisms with a brain in the behavior of plants, could we then not legitimately postulate the existence of such faculties in plants? Yes—these concepts would then have a different meaning, since their mechanisms would likely differ (no language, no brain, no representation, etc. would be involved), even though their effects may be equivalent. After all, is it not conceivable, and even logical, that evolutionary paths as radically different as the ones of plants and animals may have led to cognitive convergences with radically different causes and processes? These hypotheses are, on the whole, those entertained by plant neurobiologists and enactivist philosophers (Thompson, 2007).[16]

---

[15] Metaphors can also make communication with an audience of laypersons easier, and can thus be of use (if they are properly substantiated). Problems emerge due in part to the diffusion of scientific results by mainstream media, rather than by scientists themselves. Sometimes, media reports stick to the metaphor, which then suffers from overuse and becomes a *cliché* – while scientists, proceeding appropriately in the publication of their results, use metaphors to assist with the understanding of complex analogies.

[16] Enactivism is influenced by phenomenology and belongs to the 4E cognition movement, *i.e.* according to which cognition is enactive, embodied, embedded or extended. All views from the movement question the strict internalist view of cognition (Rowlands, 2010).

According to enactivism, any life form has a mind in a different degree determined by a continuous scale going from unicellular life to humans. As a scientific discipline, plant neurobiology starts from different bases: it postulates that plant behavior can be studied using the framework of minimal cognition (Calvo and Baluška, 2015). The discipline has naturally caused many controversies to arise, since many scientists refuse to speak of a mind without mental states such as belief or fear. But beyond the complex mental states and emotions attested in the human mind (and analogically deduced in animals), can we not conceive highly general cognitive dispositions–like desire, present in all living beings? This does not necessarily require us to make conjectures about consciousness, free will, reasoning and intentions, as some critics of plant neurobiology have sometimes claimed. Some abilities of the mind could appear at a level where biological organization is complex–and others could appear at levels characterized by lower organization, merely existing under a more rudimentary form.

Along with the understanding of plant behavior, other arguments show that the plant cognition working hypothesis is worth testing out. First, it leads to a critical shift in traditional metaphysical and scientific thought patterns. For instance, because it assumed animals had no psyche, radical behaviorism was not able to formulate hypotheses and to develop experimental apparatus to test animal cognition efficiently (Dennett, 1983; Despret, 2009). Likewise, because we thought that animals were reducible to insensible machines–or, more recently, that a piglet did not really suffer when we cut its tail–we avoided all ethical reflections on the subject. The idea that plants may express desires and suffer some prejudice which they could actively seek to avoid (without yet speaking of pain)[17] may well lead us to put new hypotheses to the test. We may then get new results and, consequently, our relations with plants could change (Calvo and Trewavas, 2020). Besides, there are a variety of ecological and environmental reasons why we should treat plants with more consideration (Stone, 1972; Hall, 2011; Hiernaux, 2018a).

Objections to Plant Cognition

Several arguments support the view that plants are more complex than machines (on this view, see Firn, 2004). This means the hypothesis of plant agency is at least theoretically legitimate. First off, some functions present in plants, although they seem machinelike, are not at all like functions in machines since they result from natural selection. Yet a process of selection and adaptation must be open to variations in behavior, and it tends to rule out inflexible, strict programming. Thus, contrastingly with machine functions, which are unmodifiable once set by their creator and heteronomous, functional autonomy in each organism remains possible. From an epistemological standpoint, saying that a plant is analogous to a machine is to account for the complexity of life using the simplicity of a mechanical model – to then say that plants are indeed as simple as machines. Picking this explanatory framework presupposes and conditions the ensuing demonstration (Calvo and Baluška, 2020). The lack of autonomy in plants cannot be validly shown in this way – which does not constitute a proof of the presence of autonomy either. Besides, as plant behavior and animal behavior are often distinguished, some arguments must provide evidence for the distinction. Where in living beings

---

[17] Pain as we feel it as animals is a subjective state. Following current knowledge, we deduce it in other species based on the presence of nerves. Thus, plant sensibility does not entail pain – although in a small group of authors (like Baluška), pain can be conceived in plants through other means, such as specific molecular receptors.

are the limits of cognition to be found? Does a mouse have a mind? What about lobsters? Ants? Slugs? Paramecia? There is no fundamental behavioral distinction between unicellular flagellate algae and "animal" unicellular organisms like protozoans.

Critics of plant neurobiology claim that plant behavior does not show the same cognitive flexibility as animal behavior–but which animals do we have in mind here? In plants, possibilities for communication and memorization are only apparent, they say. They differ in nature from those present in animals, and we are led to name the phenomena observed in plants by abusive metaphorization (Alpi *et al.*, 2007; Trewavas, 2007). These authors aim to turn potential analogies for behavior into word games with no ground in reality. That is how we would end up discussing plant intelligence. According to them, the lack of a faculty for learning during their lifespan is the strongest proof that plants can show no autonomy or flexibility (and thus no intelligence). Many biologists and philosophers–against enactivists and plant neurobiologists–think that there is no basic psychic faculty found throughout the evolution of living beings and present in the most rudimentary life forms. These authors maintain that what we encounter are rather specific emergences–like learning or consciousness–occurring at precise locations in ramifications of the phylogenetic tree. Why? Because such emergences depend on specific structures like a central nervous system (Taiz *et al.*, 2019). Since there are no such structures in plants and other non-animal organisms, their argument goes, there are no cognitive faculties either. Yet, we can criticize this argument on two grounds. First, we cannot show a necessary correlation between some cognitive faculty and some specific evolutionary structure (Le Neindre *et al.*, 2018). Second, even if we could show such correlation, it is impossible to delineate a structure like the nervous system, a fortiori in a single evolutionary branch. But still–to admit that all living beings have a psyche and experiences (yet without going as far as to say they have emotions), even in rudimentary forms, which would manifest as non-represented desires or rejections, seems too high a price to pay for many scientists.

Therefore, endorsing the hypothesis of plant cognition requires us, first, to admit that genetic or adaptationist reductionism does not provide a sufficient explanation for all of behavior, but rather that there is some minimal degree of agency in every organism. Such agency is defined by choices, desires, needs and values which are the minimal dimensions of interiority (the value system of each organism) and cannot be reduced to a strictly material dimension. Most of us in fact think about the behavior of humans (and other superior animals) from this epistemological standpoint–unless we think that even human behavior does not manifest agency, and that our experiences are entirely reducible to genes, adaptations and brain chemistry. Such a standpoint is a decisive choice, and it conditions the reasoning that follows. Finally, to say that plants have a mind and cognitive faculties is neither absurd, nor theoretically untenable. But to argue for this claim and against the opposite claim remains at present a metaphysical issue. One of the sole arguments in favor of denying plants a mind is methodological: according to the law of parsimony, when two hypotheses are equally plausible, the simplest one must prevail. Arguments in favor of plant minds, on the other hand, are pragmatic. We can also support plant cognition with a new methodological argument: to postulate some degree of cognition in all living beings should allow us to compare different species without showing any bias for groups who are allegedly the exclusive possessors of some cognitive faculties. Instead of dividing organisms in those with a mind opposite those without, could we not conceive degrees of cognitive faculties with a basis common to all organisms? For instance, could learning, consciousness and intelligence not be based on communication and memory abilities essential to all of life? And could each of these faculties not be interpreted in a variety of manners and according to a variety of scales? Beyond formal arguments, content arguments could also be of importance. To highlight some of these, we will study, evaluate and interpret some abilities displayed by plants via problems of communication, memory, learning, consciousness and intelligence. These faculties are independent to a certain extent. I have ordered them in the following sections starting with the ones whose description involves the least interiority, proceeding to the ones involving the most interiority.

Cognitive Faculties in Plants?

Communication

From a biological standpoint, communication primarily involves the transmission of information between different body parts. In vertebrates, the nervous system plays this key role. But, in plants as in any organism, information is transmitted from one body part to another through chemical, hydraulic or electrical means (Trewavas, 2014). From an ethological or biosemiotical view of behavior, however, the organic scale is the most significant (rather than the scale of its parts). We must thus investigate whether communication takes place between plant organisms. How could plant organisms communicate – given that they do not see, do not create or hear sounds, and cannot move quickly to exchange touch?

The possibility of volatile chemical communication between plants was brought up and developed – notably by the study of chemical signals exchanged underground by roots – since at least the 1980s, with the work of Rhoades (1983), and Baldwin and Schulz (1983) (Mahall and Callaway, 1991; Baluška *et al.*, 2006; Baldwin *et al.*, 2006). The principle is simple: when eaten by herbivorous predators, some plants activate an alarm signal, releasing volatile substances (such as ethylene and methyl jasmonate) which work as a warning for their other leaves, and even for other nearby plants of the same species. When they receive the chemical signal, the leaves and other plants trigger a defense reaction by producing tannins or other toxic substances which make leaves indigestible for possible predators. One of the most frequently recounted (but, paradoxically, sometimes recounted with debatable accuracy; Delattre, 2019)[18] examples of contemporary botanical literature tells how acacias have killed the kudus–a kind of antelope–of a reserve by poisoning (Hallé, 1999, p. 164–165). Today, other examples of better documented defensive reactions seem to confirm aerial communication in plants (Holopainen and Blande, 2012; Karban, Yang and Edwards, 2014). Plants can exchange information with fellow plants. But they can also communicate with other species, notably with the insects[19] and mushrooms in symbiosis with them (Bournérias and Bock, 2006, p. 191–192; Selosse, 2017). For instance, some plants react to the saliva of insects eating their leaves by sending out a chemical signal which attracts the herbivore's predator. Some plant communities even communicate via a common mycorrhizal network,[20] exchanging carbon, water and other kinds of information (Song *et al.*, 2010; Babikova *et al.*, 2013; Schulze and Mooney, 1994; Selosse *et al.*, 2007; Garbaye, 2013). But some scientists (Chamovitz, 2013;

---

[18] I thank Catherine Lenne for bringing Adrien Delattre's 2019 paper to my attention. Delattre critically reevaluates the nebulous scientific tale of the acacias and kudus. It has often been told in an approximative or variable manner by some media and popular science publications.

[19] There is another surprising anecdote concerning the way in which some acacias can "enslave" ant colonies. In such an extreme mutualism case, an acacia secretes a sweet substance which the ants use as food. This secreted substance also inhibits the production of the digestive enzymes allowing ants to digest sucrose, *i.e.* regular sugar. The ants are thus forced to stay on their food source and to protect it under pain of death by starvation: <https://www.lemonde.fr/passeurdesciences/article/2013/11/20/comment-un-arbre-mene-des-fourmis-a-l-esclavage_5998958_5470970.html> (visited on January 27, 2020).

[20] Mycorrhiza are symbiotic structures resulting from the association of a plant's roots with a mushroom.

Tassin, 2016) think this does not constitute authentic communication. According to Chamovitz, the purpose of emitting chemical messages in the air is to communicate information more rapidly to the same plant. Other plants receive this alarm signal only by accident (Heil and Silva Bueno, 2007). Since the aim of the warning signal is not to communicate with another plant, there is no intention to warn. The ecologist Jacques Tassin (2016, p. 62) likewise connects communication and intention, and concludes that plants do not communicate–reinforcing the idea that plants have no mind or cognition.

But to make intention a necessary condition of communication in order to deny plants the faculty of communication is a questionable argument. On the one hand, no proof of intention can be given since it is a philosophical concept–we can only assume its existence. Hence, to define communication as something based on intention gives us a concept which is fairly ineffective for scientific purposes. On the other hand, communication based on intention is a problem even in situations of interpersonal communication. I may well intend to communicate some information by letter, but if my letter is then lost in the mail, no communication takes place. Conversely, I may well have no intention to reveal to my neighbor a letter I wrote to vent my frustrations about him–but if my letter is somehow brought to his attention, communication will take place despite my lack of intention.

A less subjective and more scientific account defines and evaluates plant communication through its workings and its effects (Karban, 2008; Delattre, 2019). Communication is thus defined more generally as the exchange of some message between a transmitter and a receiver. It thus takes place between two (or more) individuals and requires for them to be sensible to a signal and to process the information received so as to trigger a reaction. It can also be a one-way exchange, and not necessarily involve a response. But for true communication to take place, the transmitter must be able to become a receiver, and vice versa. This "explains" how a plant which sends information to another plant which receives it communicates with it (even when it does not receive a response signal)–but how a plant which receives light information from the sun or a lamp does not communicate with sun or lamp. Communication can then not be defined by the intentions of a transmitter. By contrast, the source of information, its reception and processing, and the observable consequences it brings can provide a suitable definition.

Among plants, as well as in insects and numerous other animals, the effects of such communication can be observed. That chemical messaging may have originally served a distinct, internal purpose, to then be repurposed by natural selection for communication does not in any way undermine the definition of communication as given above. Therefore, we can understand plant communicational behavior to reach receptors in a non-intentional way.

Similarly, communication does not necessarily presuppose language. Plants and animals (and to some extent, human beings) can communicate without any representational language relying on symbolic information. Since some messages elicit a reaction in a given plant while others do not, we can reasonably think that some messages have a signification while others do not. Signification does not always involve representations or other mental states (like intentions)–and it could well be some elementary indicator of mind as enactivists or plant neurobiologists conceive it.

A strictly adaptationist reading does away with the experience of signification in plants. It interprets the phenomena we described as an evolutionary optimization of information transfer. We are thus led back to our initial alternative: whether or not to opt for methodological reductionism–knowing that unresolved issues are entailed by the latter.

Memory and Learning

Drawing on 20th-century classical adaptationist arguments, the inquiry will now return to the idea that plants are
programmed by the results of natural selection. These would
account for the optimized reactions of plant organisms to
their environment without relying on any true agency on
their part. In this view, variations in the behavior observed
in plants never result from an interpretation of what has
signification for them among the objects and events in their
environments–but result rather from automatic reactions.
Plants could only inherit behavioral innovations–and not
adapt their behavior during their own lifetime. For only
learning can account for such flexibility and authentic agency
in higher organisms. Indeed, whereas communication could
be understood to be solely the result of natural selection,
learning is fundamental to behavioral variability. According
to this view, an organism could communicate, and yet be
unable to integrate new kinds of signals, or to interpret them
differently during its lifetime.

It therefore becomes crucial to know whether plants can
learn. If they can, the adaptationist account can no longer
apply to all kinds of innovative behavior which appear to be
directed at a goal. Indeed, learning involves a change in the
internal state of an organism which thus gains control over
the effects of a behavior or modifies them. More precisely,
following the definition in Okano *et al.* (2000), learning is
a process of information memorization which enables adaptative changes in an organism's behavior in response to its
experience. Learning is then closely linked to memory. And
asking if plants can learn requires us to first inquire into the
existence of plant memory.

Memory

To show that memory exists, we must observe a response to a
stimulus delayed beyond the reaction time of reflexes or mere

irritability. This entails some notion of time, which can even fall into line with basic consciousness:

> [...] [the] leaves of some plants, such as *Lavatera cretica*, can anticipate the direction of sunrise, even after they have been prevented from solar tracking for several days. The combination of memory and anticipation is consistent with the phenomenological description of time as the retention of a past "now-moment" and the projection into a future "now-moment" by a conscious subject. The sense of place remains incomplete without this, its experiential temporal dimension. (Marder, 2012, p. 3)

Without going so far as to invoke parallels with a conscious subject, how can we understand memory from a scientific perspective? The biologist Michel Thellier (2015) tells us that plant memory has three functions: storage (when information is memorized), recall (when information is reactivated) and inhibition (when the information to be reactivated is selected or not). Is such a memory similar to the memory of humans and animals?

According to experiments, plants can store, reactivate and even inherit information at a cellular level. For instance, the botanist Jean-Marie Pelt (1996, p. 170–173) cites an experiment performed on bryonies, a kind of climbing plants, where bryonies react to aggressions against one of their internodes (in this case, friction) with increased lignification. If we collect cells from the internode while it defends itself with lignification and cultivate them in a culture medium, then collect cells from this first culture medium and cultivate them in a new culture medium and so on, we will find lignification to be higher than normal in the cultivated cells until the fourth generation. It is as if the plant cells "remembered" the aggression signal which they inherited. They keep track of it via cellular information (Bourgeade *et al.*, 1989). The perpetuation of inhibitory messages acquired in reaction to stresses has also been found in other plants (Trewavas, 2014, p. 211–219). Such intergenerational memory likely comes from epigenetic modifications.

But it is sometimes thought that memory must necessarily be connected to the presence of strong (*i.e.* semantic) information—and that we cannot infer memory from traces of weak (*i.e.* epigenetic, non-semantic) information stored in plants (see Maher, 2017, p. 85–88). "True" memory, in short, would minimally require semantic memory. Plants, however, store and recall weak information which, differently from semantic information, cannot be false since it is neither formulated, nor represented. Even if plants store and restitute information, they would not thus truly display memory—because rocks too can store heat and restitute it. They would only be keeping traces (Tassin, 2016, p. 120) just as the ground keeps traces of footprints.[21] According to such reasoning, mental representations based on semantic information remain the keystone of the mind.

On another note, if we accept that there is a mind without representation, it becomes entirely possible to also accept that retention of weak information provides evidence for memory. This distinction is similar to the way ancient and medieval philosophers distinguished between *anamnesis* and *mneme*. For Aristotle, some animals "high enough in the chain of living things" possess *mneme*: a direct, spontaneous or passive memory. *Mneme* opposes *anamnesis*, which is "exclusive to man because it supposes recall, consciousness, memory efforts" (Simondon, 2004, p. 44). *Anamnesis* is the faculty of conscious recall and memorization. The English language does not distinguish these two kinds of memory. But when it comes to plant behavior, the distinction is paramount. It is indeed crucial that we detach ourselves from the notion of consciousness—a notion explicitly rejected by most botanists—in order to pursue a philosophical inquiry

---

[21] The trace left by a fossilized leaf stored in a rock is neither true nor false. It is simply there (or not). On the contrary, the mental representation I express with the proposition "There is a trace of a fossilized leaf stored in this rock" can be semantically true or false, since, in the process of representing the fossilized leaf, I can err in the perception and in the treatment of information.

into plant memory. Against Aristotle (and presumably Simondon), recent science taught us that all living beings have *mneme*–even plants–and that numerous vertebrate species also have *anamnesis*.

Overlooking the distinction between these two kinds of memory can lead us to make sweeping statements. Thus, the ecologist Jacques Tassin, while he acknowledges the interest of experiments on plant memory, concludes that there is no such memory (Tassin, 2016, p. 77–95):

> [Plants] no doubt integrate a past event in their response to a new event. But do they remember? A more sober thought seems more appropriate: for fifteen days, seedlings keep track of wounds as information which can interfere with other information. (Tassin, 2016, p. 85).

Tassin's main claim is that we may mistake a mere trace for true memory, which consists in "remembering". But memory cannot be entirely assimilated to a remembering which involves representation and consciousness. In humans, three kinds of memory exist (Tulving, 1985; Sternberg and Sternberg, 2017; Michaelian and Sutton, 2017). Episodic memory concerns events and memories. Semantic memory is for concepts and information. Procedural memory is used for learning and allows us to record series of operations. The storage, recall and inhibition functions displayed by plants rather point to a form of procedural memory. The conscious recall (*i.e. anamnesis*) Tassin mentions is a feature of the episodic (and semantic) memory found in humans and in some animals.[22] This conception of memory based on consciousness and representation is the one our modern philosophical tradition preserved almost exclusively. The ancient and medieval distinctions of *mneme* and *anamnesis* have been expunged for the benefit of a definition of memory as a "psychic function where a past state of consciousness

---

[22] According to Tulving (1985), there are degrees in these kinds of memories which correspond to degrees of consciousness emergence in living beings.

is reproduced in such a way that the subject is aware of the reproduction" (Lalande, 1996, p. 606–607). Such a subjectivist approach may have serious impacts on the interpretation of plant behavior as a whole (communication, learning, intelligence), as Tassin's position shows:

> Intelligence is a capacity for one to adjust their behavior during one's lifetime. It supposes an ability for choice connected to a learning faculty which only a truly integrative memory can warrant. Such memory, because it knows how to remember, also knows how to face new situations. But nothing provides evidence for it in plants, where all we find are more or less persistent traces. No–plants are certainly not intelligent. They do not memorize anything, neither do they foresee anything, in keeping with Hegel's account of a being condemned to immediacy (Tassin, 2016, p. 120).

Once again, this speaks to the influence of our modern Western philosophical tradition, in which, in order to have a mind, one must be able to represent things via mental acts. But is the ability to record information in order to reuse it later in a way that serves a plant's interest not precisely a more objective kind of memory? For this is the first and most general of the five senses given for "*mémoire*" in the French dictionary Larousse: "Biological and psychical activity to store, preserve and restitute information" (2002, p. 642). Plants will then express more than a simple trace. A rock keeps a trace of any shock that may have hit it, yet it is unable to inhibit or to reactivate such trace later–as any organism could when adapting its behavior to a different or similar situation. A rock can restitute stored information (for instance, by rolling, cooling down or breaking), but the information stored in a rock is not integrated or regulated by any finality and therefore by any value system. By contrast, long-day plants, for instance, flower when they detect a red-light stimulus and keep it in their memory as cellular information, but they can also inhibit their flowering if they then perceive a new far-red light stimulus (Chamovitz, 2013, p. 33; see also Raven *et al.*, 2013, p. 668–673, for a similar example involving seed germination). There is thus retention, but also treatment and coordinated management of information according to a plant's interest (in this case, the optimal moment for flowering). Memory is then indeed a biological and psychical activity: plants display memory, but not stones (even if plants and stones both record weak information). Memory is a global process based on sensibility. The storage, recall and inhibition of information by an organism depend on what has signification or value for it. Moreover, learning can follow from memory – and stones, while they do record weak information, cannot ever learn.

Learning
We often a priori take learning to be a typically animal faculty. This is because learning allows for behavioral flexibility, which seems to set it against strictly genetic or adaptationist explanations and could thus suggest the presence of cognition or mind. According to a traditional view, as represented by Staddon (1983, p. 395) for instance, there would even be a correlation between learning and the size and lifespan of species. So-called simple organisms could not learn because they lack the required neuronal mechanisms – but also because they do not live long enough to experience different niches and use what they would have learned. This is how one would argue that learning belongs exclusively to superior animals.

But the argument is open to several critiques. First, it is not the (neuronal) structure which accounts for learning, but the result of learning – *i.e.* that information is stored and reusable, in such a way as to modify behavior accordingly. Second, even if we took size and lifespan into account, we should conclude that some plants – like trees – learn, since their size and lifespan are often significant. And it is true that perennial plants do not move about in their environment, but their lifespan can require them to adapt to new niches over the course of their existence. Third, in any case, experiments have shown that all organisms learn – even when they are very small and have a very short lifespan, like bacteria (Kawecki, 2010).

Yet, that learning may be possible in plants has generated vehement criticism – more so than memory. The journalist Michael Pollan relates the following anecdote from his participation in a conference on plant biology where the biologist Monica Gagliano presented the results of her experiments on the subject:

> On my way out of the lecture hall, I bumped into Fred Sack, a prominent botanist at the University of British Columbia. I asked him what he thought of Gagliano's presentation. "Bullshit," he replied. He explained that the word "learning" implied a brain and should be reserved for animals: "Animals can exhibit learning, but plants evolve adaptations." He was making a distinction between behavioral changes that occur within the lifetime of an organism and those which arise across generations. At lunch, I sat with a Russian scientist, who was equally dismissive. "It's not learning," he said. "So there's nothing to discuss." (Pollan, 2013)

This leads us back to one of the *idées fixes* of the controversy: plants are allegedly unable to adapt on an individual level during their lifetime (*i.e.* unable to learn) – rather, they adapt exclusively through natural selection. Are we then not tackling some kind of biological dogma?

Either way, asking the question encourages us to consider the ethological problem of learning by habituation in plants. Habituation has two forms: habituation strictly speaking and sensitization. Habituation occurs when an organism learns to ignore a potentially negative, but harmless stimulus – or to minimize its reaction to it. Sensitization, on the other hand, occurs when an organism anticipates a harmful stimulus. Are these forms of learning to be found in plants?

Recent discoveries suggest that plants, like animals, possess general learning mechanisms:

> The plants which have been stimulated change their response to a new administration of the same (or, sometimes, of a different)

stimulus. This is what we can call a memory of the "learning" type. It may seem incongruous to use the word "learning" when we talk about plants. But, today, we do use this word for softwares, robots and for some machines, so why not for plants? (Thellier, 2015, p. 45–46).

There appears to be a correlation between the learning modalities studied in plants and the epigenetic modifications which induce a change in internal state and produce a new behavior which is better adapted to a recurring situation. Consider the *Arabidopsis thaliana* model plant. It reacts to a cold shock with an increase in cytosolic calcium[23]–but the reaction is mitigated if the arabidopsis has undergone a prolonged or repeated exposure to cold beforehand. This is only one example among the numerous types reported (Thellier, 2015, p. 61). What takes place is thus a cellular or corporeal learning which is somewhat different from our anthropological or animal understanding of this faculty.

Based on these physiological experiments, Gagliano *et al.* (2014) have wondered whether it was possible to train sensitive plants (*Mimosa pudica*) so as to induce a directly observable behavioral reaction resulting from learning. The sensitive plant is an ideal candidate for learning by habituation, since it can fold its leaflets following mechanical disturbance (*i.e.* thigmotropism) interpreted as harmful (because it is a priori associated with the presence of some predator). But this mechanism comes at a cost: a 60% decrease in the photosynthesis yield (this being added to the energy cost of the reaction itself). The working hypothesis is that it is in the best interest of sensitive plants to learn not to fold their leaflets (or to fold them for a shorter time) when reacting to a harmless touch in order to avoid an unnecessary loss in photosynthesis yield. The hypothesis of such habituation was tested through

---

[23] The calcium ions (Ca^{2+}) found in the liquid component of the cellular cytoplasm (cytosol) of plants play an important role in the biological response to environmental stimuli through the activation of certain genes (Xiong *et al.*, 2006).

a clever dispositive which produced regular drops in a group of potted sensitive plants, but not to a control group. The result? A significant habituation was observed in the experimental group: after a series of drops, the leaflets remained closed for less time. But–one may ask–is this not simply due to exhaustion or lack of energy in the experimental group of plants? The experimenters refute this possibility in two ways. On the one hand, the sensitive plants who have developed habituation to drops will immediately recover their pre-experimental reaction time if the stimulus changes–for instance, if they are touched rather than dropped. On the other hand, the new reaction is indeed the result of learning, because it is kept in a plant's memory for up to 28 days after habituation has been developed, with no new stimulation being provided (Gagliano *et al.*, 2014). Even though it operates at a cellular level, this kind of behavior involves memory and learning, and its effects are similar to those observed in an animal with a nervous system.

Philosophically speaking, a lack of learning in plants would have argued in favor of a distinction between plant behavior and the behavior of animals who possess a mind and cognitive faculties. But both animals and plants (or at least, the species tested) can develop habituation in addition to a trial-and-error phenomenon, which was previously noticed in plants. Darwin had observed some climbing plants testing their way up until they find an optimal structure for attachment. They can even let go of an inappropriate structure and find a new one if their assessment is mistaken–for instance, if it is too smooth, or if it is another tendril. Likewise, the dodder (*Cuscuta europaea*)–a parasitic plant – can change host in case of error (Kelly, 1992; Trewavas, 2014, chap. 9). Recent experiments demonstrate that the nutation movement of peas is neither linear nor purely endogenous because it is influenced by the presence of a distant pole. Peas are seen to detect the support before touching it and to adapt their movements to its presence in a complex way, suggesting that the biological control of nutation is guided by relevant environmental factor like the pole (Raja *et al.*, 2020). Moreover peas' tendrils adapt their aperture to the pole thickness, before touching it, by modifying their velocity (Guerra *et al.*, 2019; Simonetti *et al.* 2021)[24]. The ability for discrimination and adaptation to one's environment by trial and error, through choices and decisions would thus exist in all plants−as shown by niche research and optimization (Turkington and Harper, 1979; Trewavas, 2003). Plants would then not only display passive habituation, but also active learning through the exploration of their environment.

Moreover, plant behavior can no longer be reduced to mere physical causality, as memory could be, to a certain extent−for nowhere in the physical world is there something like learning. The integration and restitution of weak information, by contrast, can be found in stone, but it is no proof of true memory. This no doubt explains why learning terminology generates greater deadlock and affronts than memory terminology: Gagliano's work has been rejected for publication in several journals due to the terminology it uses, although the reliability of the data it reported has not necessarily been questioned.

Plants may well display learning by habituation−but perhaps is learning by sensitization found only in animals. But a recent experiment on peas by Gagliano *et al.* (2016) nevertheless shows a kind of learning by association in plants.[25] After being trained to associate a source of light with a fan draft, peas will preferentially orient their stems toward the draft, even in the absence of light. The association of draft and light source is not observed in the control group. Traditionally exclusive to animals, this ability for associative learning

---

[24] Although the literature observed and experimented these phenomena, their underlying mechanisms remain unknown, especially for the distant detection of the pole.

[25] I am grateful to Bruno Moulia for drawing my attention to this experiment.

brings further support to cognitive faculties in plants. Some scientists, however, have questioned the methodology of the experiment and the possibility to replicate its results (Taiz *et al.*, 2019; Markel, 2020). Yet, to this day, it is not memory or learning which has most drawn the attention – especially in the general public – where the frontiers of mind are concerned: it is consciousness and intelligence.

Consciousness

> The reason that controversy can surround the application of the word "intention" to the behaviour of organisms surely arises from the fact that human intention commonly involves conscious action, and consciousness is judged on the basis that only very similar organisms to ourselves can be conscious (Trewavas, 2014, p. 90).

If we admit consciousness as a necessary condition of mind, then plants, like most animals, seem deprived of it. For the traditional definition of consciousness calls it a faculty which "supposes a clear opposition between what knows and what is known, and an analysis of the object of such knowledge" (Lalande, 1996, "Consciousness", p. 174). What we have here is a kind of introspective consciousness of oneself, in which one is able to recognize oneself as a thinking or perceiving subject. It is hard to imagine a plant being able to express such a Cartesian ego. For a long time, such introspection has been held to be exclusively human. Yet experiments – like those based on the mirror test, among others – have shown that birds, primates and other mammals recognize themselves as subjects of perception (Despret, 2014, p. 169–179).[26] More broadly, a recent collective expertise on forms of consciousness in animals, based on their cognitive aptitudes and neurological mechanisms, has shown that numerous vertebrates (but not only them)

---

[26] Experiments conducted on primates have also brought to light that they had intentions and could attribute intentions to fellow primates (Despret, 2014, p. 133–144).

possess access consciousness,[27] introspective and social self-consciousness and consciousness of the state of their knowledge (Le Neindre *et al.*, 2018). A priori, plants do not possess reflexive consciousness (which allows knowledge of one's emotions, one's role in a social organization, and one's perceptions, knowledge or actions). For such faculty requires mental representation. Does this mean we only need to bring up consciousness to establish the definitive boundaries of a psychic level of animal behavior, and thus distinguish it from a mere biological level of behavior?

In fact, philosophy acknowledges different kinds of consciousness (which are, to some degree, interdependent). Apart from reflexive consciousness, there is a spontaneous awareness, *i.e.* an immediate consciousness of one's environment (Lalande, 1996, *ibid.*). Some scientists have argued for the existence of such awareness in all living beings—therefore including plants (Margulis and Sagan, 1995; Chamovitz, 2013; Trewavas, 2014, chap. 25). Plants are thus aware of the type of light and contact, of gravity and of the chemical signals they receive, of their past experiences and of the conditions of anterior physiological changes (Chamovitz, 2013, p. 166).

But such "consciousness" of one's surroundings in fact amounts to the treatment of information coming from the environment (or, from the standpoint of memory to the rekindling of past information). This explains how this kind of consciousness can appear in any sensitive organism. Are there modalities of awareness more specific than such immediate consciousness in plants?

Certain kinds of self-recognition (like corporeal or "social" consciousness), involving more than the mere immediate awareness of one's environment, leads one to ask whether

---

[27] Access consciousness is the capacity to use representations in behavior, whether such behavior is to relate to the environment or to communicate (Proust, 2003, p. 164).

some intermediate form of consciousness may be involved. Such faculties manifest in various ways, every one of which requires one to minimally possess a value system distinguishing the plant's self from its environment.

Experiments have shown that numerous plant species do not react in the same way when one of their roots comes in contact with another of their own roots and when it comes in contact with a root from another species (Sultan, 2015, p. 54). In the latter case, growth is inhibited and the root distances itself from its neighbor, leaving a patch of unoccupied soil between itself and its rival. In the former case, no inhibition takes place (Mahall and Callaway, 1991, 1992, 1996). These experiments then point to a form of corporeal consciousness, and even to some simple social consciousness regarding related plants. Experiments also indicate that two plants cloned from the same individual acquire a foreign character toward each other after being separated for some time (for instance, this occurs after 60 days with two pea clones) (Gruntmann and Novoplansky, 2004). The hypothesis of such self-recognition is based on integrated physiological mechanisms which would no longer be effective once the clone has undergone too many epigenetic changes, or too many changes in the mechanism controlling recognition (Trewavas, 2014, p. 188–189). This is then in favor of some kind of plant autonomy manifest in a plant's own behavioral individuality (which can be innate or acquired). Furthermore, biological dynamics of competition and cooperation also imply some form of self-recognition. In symbiosis cases, immunology shows that the limits of the "self" are not necessarily those of the genetic organisms which can constitute a functional entity with a multispecific identity (Pradeu, 2010; Selosse, 2017). Proprioception is another example of plants showing some knowledge of their own body (Bastien *et al.*, 2013; Dumais, 2013). Proprioception is the capacity to know where the different parts of one's body are relative to each other without seeing them or touching them. Drunkenness alters this faculty (touching one's nose with one's eyes closed becomes more difficult). Proprioception is hard to grasp, even in animals, because it depends on the coordination of different body parts via the internal ear and specialized proprioceptive nerves. This faculty encompasses both the balance signals of limbs at rest and the dynamic coordination of limbs in motion. Plants also possess this static and dynamic "consciousness" of their body (Chamovitz, 2013, chap. 5). It allows trees to reorient their growth to compensate for an imbalance (for instance, due to too important a growth on one side or to damages caused by lightning) and keeps them from collapsing under their own weight. All these elements indicate that any spontaneous awareness already involves some form of self-consciousness. Plants can then not be considered to live in a purely non-differentiated and immediate continuity with their environment. Some minimal framework of alterity and self then appears essential in plants, contrarily to what the philosopher Michael Marder suggests (2013a).

Indeed, in plants, the possibility of discrimination between self and other, and even more finely between the self and the other as environment (*i.e.* as a resource or obstacle), or the other as another self (*i.e.* as a member of the same species or reproduction partner), or the other as other than oneself (*i.e.* as a member of another species) – with the refinements that the last option supposes (since such another can be neutral, or a predator or cooperator) – does exist. Of course, plants do not conceive of these various modes of interaction, but, through their reactions, their behavior shows some discrimination between these modes. Possessing a kind of minimal self-awareness which requires no intention or reflexivity, plants manage to effectively resolve the problems they encounter in the course of their life. They do so by interacting with each other and by adapting to their milieu thanks to memory and learning. Some authors incidentally take such aptitudes to fit the very definition of intelligence.

Intelligence

So–are plants intelligent? Beyond the controversies, an implicit conception of life and plant life surfaces. For is intelligence not the sign of rationality, cognition–and mind?

In the historical review above, we have seen how the Western philosophical tradition tends to deny plants any kind of intelligence, mainly because this faculty is closely related to movement and subjectivity. This has remained the leading standpoint among contemporary scientists and philosophers who generally take intelligence to be a conceptual, rational activity, or then an activity involving conscious choices (Lalande, 1996, p. 524–525). One of the philosophers who depicted plants along these lines in the clearest, fiercest manner is arguably G.W.F. Hegel (1771–1831). In his *Philosophy of Nature*, plants are conceived to be in a strictly immediate relation with the outside world due to their supposed lack of self, which entails a lack of a subjective relation to themselves. Such a lack is accounted for by the fact that plants cannot move, nor can they cut themselves off from soil, water or light (Hegel, 2004). In this view, plants are trapped in an infinite process of absorption and rejection which they cannot interrupt or refuse–unlike animals (Miller, 2002, p. 138).

However, the numerous scientific experiments we have mentioned now do not appear to fit well with the conception of plants as passive.[28] Plants can interrupt some of their metabolic processes, close their stomata to regulate water loss, make their leaves parallel or perpendicular to sun rays in order to receive more light or to avoid burns, etc. The main argument of Firn (2004) against the intelligence of plants displaying such behaviors is to invoke their lack of individuality. The reaction of a plant as a whole–a reaction which appears intelligent–would be nothing more than the

---

[28] In fact, this inconsistency can be traced back to the 19th century (at the latest) and the first physiological experiments on plants (Hiernaux, 2019).

congruent sum of the genetically programmed activity of each one of its parts. In contrast, Trewavas (2004, 2014) has argued for plant intelligence by showing that–despite the fact that plants are less centralized than animals–some of their reactions are a proof of internal communication and thus of a true integration of the whole plant as a behavioral unit. To oppose intelligence and programmed, incurred activities implies the identification of intelligence with consciousness, including conscious representations, intentions and will. Yet intelligence goes beyond consciousness–a fact which accounts for artificial intelligence terminology, even though the activities of software are programmed and even though software does not display consciousness. However, the authentic intelligence of a living being cannot be thoroughly programmed the way a machine can. Indeed, even if some behaviors are based on activities mostly programmed, intelligent behavior must remain open to some degree of variability. The cases of learning and trial-and-error described in plants precisely involve behavioral variability and accordingly allow us to postulate a kind of minimal intelligence in plants.

The possibility of a behavior is the result of natural selection at the level of behavioral structure–but it does not determine all effective behavioral activities of an individual which such a structure makes possible. I then suggest the following: to consider intelligence to be limited to such effective behavioral activity, rather than to extend it to all effectual programming and all genetically acquired adaptation. For this working hypothesis allows for the dissolution of ambiguities in several controversies.

However, such a view calls for a more "pragmatic" and situated conception of intelligence to be applied to plants. Trewavas (2002) thus borrows the definition of intelligence as possession of an "adaptively variable behaviour during the life of the individual" (Stenhouse, 1974). His *Plant Behaviour and Intelligence* offers a more precise definition:

Intelligence measures an agent's ability to achieve goals in a wide range of environments. Features such as the ability to learn and adapt, or to understand, are implicit in the above definition as these capacities enable an agent to succeed in a wide range of environments (Trewavas, 2014, p. 195, drawing on Legg and Hunter, 2007).

Here, intelligence is assessed according to its consequences—and not according to the intentions or conscious will which lead up to its actions. Plant intelligence thus overlaps in part with phenotypic plasticity. For instance, the well-known adaptation of leaf structure in the common water-crowfoot depending on its leaves being underwater or floating. Trewavas (2014, p. 84–85) also recounts experiments on the capacity of plants to discriminate between situations (between several types of juxtaposed soil, or several supports for vines) and to make a choice—which appears optimal to us—after a trial-and-error process.

Taken in that sense, intelligence can also be argued for within a continuist view of life forms and belong to all organisms, even unicellular organisms and plants. Intelligence is thus a faculty which makes the behavioral activities of all living beings measurable via their adaptation to their environment. However, could one not object that any biological behavior thus necessarily becomes intelligent? Intelligence would then lose its appeal.

Such an argument is based precisely on the confusion between behavioral structure and its effective manifestation. Likewise, debates on plant intelligence often conflate the adaptation of species or populations with the adaptation of individuals. For instance, Mancuso considers the co-evolution taking place between a plant secreting nectar and the ants feeding on it—and, in turn, defending the plant—to be intelligence (2015, p. 24). By contrast, Trewavas (2014) explicitly distinguishes the two adaptation levels we have mentioned and uses the term intelligence exclusively for the behavioral adaptations of organisms. Even if these two types of adaptation increase an organism's chances to survive and reproduce, the adaptive plasticity of behavior contrasts with the heritable adaptation of a phenotype, like the wings of birds being well adapted to flight (Trewavas, 2014, p. 194).

Consider this example: it is not because a hen has the required behavioral resources to escape a fox (running, or flying away, for instance) that it will necessarily escape it. In truth, a behavior can be intelligent or not if it involves a possibility of error or malfunction. Theoretically, there are then indeed some behaviors which are intelligent and some which are not, even if they both presuppose the structural result of selection and adaptation. If the experiment on associative learning by Gagliano *et al.* (2016) is eventually confirmed, it would provide a strikingly convincing case of plant intelligence since all individuals tested do not incorporate the conditioning they received as efficiently in their subsequent behavior. This kind of intelligence is then an individual adaptation to a novel situation.

In this regard, the individual which adapts best, no matter the means, is also the most intelligent, which may be counterintuitive starting from the more standard, subjective conception of intelligence. Indeed, a classical conception of intelligence (of human intelligence, for instance) readily considers that only some means of adaptation show intelligence. In this way, adapting to a situation with cleverness is taken to display intelligence, while an equivalent adaptation performed by force appears less intelligent. But in the biological debates on intelligence, all the means of adaptation available to an organism are equivalent. Intelligence conceived within the evolutionist framework of fitness is therefore far from our intuitive or philosophical notion of intelligence.

Intelligence then concerns biological aspects because it takes into account any variation in an individual's aptitude for problem-solving–but that does not mean that psychic and conceptual aspects are reduced to biology. Even if plants possess a kind of intelligence common to all living beings, such intelligence is not to be conflated with animal or human intelligence. Thus, as fixed beings which are not centralized by a nervous system or brain, plants express their behavior – and therefore their intelligence – mostly through the plasticity of their body (growth, change in state, resilience). They do not reach levels of abstraction such as mental representation (and language) or reflexive consciousness, which are likely the earmark of the organisms close to us. Some, like Mancuso, thus see in plants a kind of intelligence similar to the decentralized intelligence of insect colonies which could be of use to us as more specific sources of inspiration, for instance in the development of artificial intelligence.

In conclusion, an organism which works well and which meets its needs intelligently is also an organism that is more likely to survive and reproduce. Biological intelligence is thus quite removed from the notion of a rational faculty, intrinsic to an individual. Intelligence resides in the adaptation relation between an organism and its environment (Calvo, 2016). In this regard, it is never an internal, abstract faculty – it should rather always be considered in its relation with a given environmental and corporeal situation. Aptitudes like communication and learning testify to this relationality. Affordance theory, biosemiotics (Witzany, 2008) and behavioral ecology applied to plants tend toward the same conclusion (Gagliano, 2015). Indeed, just as biosemiotics considers that an organism's milieu does not appear to it in a neutral way, behavioral ecology considers that the environment contains action potentialities relative to each species – which potentialities are thus not intrinsic to it (*i.e.* affordances).

Biosemiotics and Plant Behavior

The leading framework for the scientific interpretation of plant behavior has been covered above. I now want to put forward a non-reductionist, more philosophical reading, starting from a critical rework of concepts from biosemiotics. What would the world of a plant be? What significations–in the minimal sense mentioned–could a plant give to its milieu? Can we take a plant to be a kind of knowing subject with a kind of interiority?

Introduction

According to Drouin (2008, p. 195), who draws on Hallé, Aristotle and Bergson, if plants benefit from a novel, redefined and decentralized form of individuality, "it however seems obvious to deny it any subjectivity". But how would that be obvious? If we follow Descola (2005), reluctance to subjectivize plants is grounded in (naturalist) modern Western ontology:

> In modern ideology, the discontinuity between humans and other beings originates in a conception of human interiority as doubly subjective: self-consciousness constitutes subjectivity, subjectivity allows moral autonomy, moral autonomy is the basis for the two properties of the subject as an individual having rights and obligations toward a community of equals–responsibility and freedom. Being traditionally depicted as lacking these properties, plants and animals are then excluded from civic life: it is impossible to have political or economic relations with them, since they have no subject status. But such subordination of non-human beings to the decrees of an imperial humanity is increasingly contested by legal and moral theorists who work toward novel environmental ethics–no longer burdened by the preconceptions of Kantian humanism (Descola, 2005, p. 268).

Even in the ethical theories which include animals, like Singer's (1975), principles of naturalist ontology predominate. Animals are the objects of the moral care of humans, but they remain subordinate to them since they are not considered to be autonomous subjects. What happens to plants in this framework?

> Lacking sensibility, plants and abiotic elements of the environment remain sentenced to the mechanical and impersonal

fate which naturalism previously bestowed on all non-human
beings (Descola, 2005, p. 271).

Consequently, if animals are not autonomous subjects, they at least reach a kind of heteronomous subjectivity while plants remain among objects, taken together with the "abiotic elements" of the mineral environment. Yet, opposite the standpoint of the naturalist ontology recounted in Descola, plants do have sensibility. Should this simple fact not suffice to distinguish them from inorganic matter and from the objects naturalism takes them to be? At least, this is what pre-modern, non-Western animist ontologies recommend by recognizing, for instance, animals, but also plants as non-human persons (Hall, 2011).

Instead of this polarized debate, could we not also consider – in a more situated way – a "subjectivity" custom-made for plants based on what we know about them? This requires a disconnection between subjectivity and person or human. Indeed, faculties generally associated with human subjects like reflexive consciousness, representation or conceptual intelligence cannot apply to plants. Yet, beyond all the behaviors of an individual plant (such individuality being sometimes decentralized), there remains some cohesive behavioral unit. Is some subjectivity not required to secure such unity? Taken in a broader sense, such subjectivity can be understood – more fundamentally – as what would define the specific relation of plants with time and space. That relation would provide a starting point to account for specific behaviors. Concerning the connection between the ethological and philosophical dimensions of subjectivity, I draw mainly on Jakob von Uexküll's (1864–1944) *A Foray into the Worlds of Animals and Humans* (2010). Uexküll is one of the founders of animal ethology, but he is also a philosopher and influenced many later thinkers (such as Heidegger, Merleau-Ponty and, more recently, Augustin Berque and his mesology) by putting forward a new conception of (animal) subjectivity (Pieron, 2009, 2010).

Uexküll's ethology is of particular interest because it allows us to test the hypothesis of plant subjectivity. Is it as absurd as modern naturalist ontology takes it to be? A mere zoomorphic transposition would completely miss the point here. The task is rather to develop a critical interpretation of plant subjectivity – including the discrepancies and differences involved – in a continuum with the scientific perspective developed in previous chapters.

Animal and Human Milieu

The starting point of Uexküll's philosophy is the rejection of animal mechanism. Living beings use means of perception – *i.e.* perception-tools – and means of action – *i.e.* effect-tools – behind which a subject is to be found. Like humans, animals are no mere machines, because, behind their tools and perceptions, subjectivity can be found. Machines, on the other hand, would be the same as their means of perception and action. Following Uexküll, any animal is a subject with a perception world (the sensible world, *Merkwelt*) and an effect world (*Wirkwelt*) which together constitute its milieu (the *Umwelt*, a species' own world). For any organism in a given species, the potential of action, of course, depends on its potential of perception.

Uexküll has thus provided a theory to account for the subjectivity of animals then called "superior" – but also of invertebrate animals – in relation with their milieu (*Umwelt*). Indeed, his paradigmatic example is the tick. My method draws on his analyses to extend ethological hypotheses to plants.[29] For ticks' capacity to react to stimuli seems extremely limited, since their potential of perception is limited: they can perceive light, heat and the smell of butyric acid released by their mammal hosts. According to the scientific results

---

[29]   Emanuele Coccia condemns any possible plant ethology drawing on Uexküll – the reason being that plants have no sense organ dedicated to their relation with the world (2016, p. 57–59).

mentioned above, plants are sensitive and react to at least as many stimuli (whence their behavior) as ticks do. We can thus attribute to plants a perception world and an effect world, even if Uexküll does not.

Some scientists who study plant behavior share Uexküll's criticism of mechanism. However, for Uexküll, such criticism goes hand in hand with the subject-object dualism, in which subject and object are mutually exclusive theoretical alternatives:

> Is the tick a machine or a machine operator? Is it a mere object or a subject? (Uexküll, 2010, p. 45).

Whether they want it or not, plant biology scholars who use classical ethology to understand plant organisms find themselves within naturalist subject-object dualism, as it has been "animalized" by Uexküll. For, if animals have been pushed back into the margins of modern Western thought, this means that all "other" living beings–including plants–have been pushed back even farther into "the zone of absolute obscurity undetectable on the radars of our conceptuality" (Marder, 2013a, p. 2). Thus, plants have been reduced nearly entirely to scientific objects, as if their being was unsuitable for philosophical thought.

Centralized Subjectivity, Non-Centralized Subjectivity

In *A Foray into the Worlds of Animals and Humans*, Uexküll suggests an important distinction between simple animals whose reflexes are not centralized (like jellyfish and sea urchins) and other animals whose reflexes are centralized (so-called superior animals). In the former animals, every reflex arc is autonomous and independent–which leads Uexküll to say: "When a dog runs, the animal moves its legs. When a sea urchin runs, its legs move the animal" (2010, p. 76). The result in a sea urchin is that the stimuli perceived via different reflex arches (*i.e.* via different organ types) remain isolated–they are not centralized in one perceptual object. Another hypothesis is that, in centralized animals, subjectivity dominates action which it controls, while in decentralized animals, subjectivity seems to result from action.

Are plants more like sea urchins, or more like superior animals? The answer likely depends on the kind of plant organism concerned. Very basic plants like unicellular algae are rather like the sea urchins, while flowering plants–which are involved in most of the experiments we have mentioned–are somewhat like vertebrates in some respects. Indeed, flowering plants react to some stimuli like centralized animals do. For instance, only one tree branch needs to undergo vernalization through cold for the reaction signal to be sent to the whole organism. If a parasite attacks the leaf of a plant in one specific, localized spot, the defense reaction (*i.e.* toxin production) can be triggered in the entire foliage to ensure the plant's protection against other eventual attacks. Likewise, each leaf possesses phytochromes able to detect light and a sole leaf exposed to the appropriate type of light can trigger flowering in the entire plant. However, this kind of centralization is not the same as nervous or cerebral centralization. Indeed, aspects of the morphology and autonomy of plant parts suggest a weaker individuality. A fundamental feature of centralization is nonetheless found in plants, *i.e.* communication of information in all plant parts. To understand what a plant *Umwelt* would be, it is not enough to oppose a vertebrate centralized animal model to an invertebrate non-centralized one, nor to study the mechanisms plants use in their relations with their milieu. Rather, their "subjectivity" may depend on the nature of functions and on the signification of these relations.

Impoverished Milieus and Certain Milieus

For Uexküll, a milieu's complexity or simplicity depends on the specific organism to perceive it and relates to the complexity of that organism's world of perception. For instance, ticks can only react to three different properties in a definite order. These properties cause excitation and reaction: the tick's milieu is impoverished. However, the milieu's poverty is not a value judgment,[30] because the simplicity of a milieu increases certitude in animals when they act accordingly, and the degree of certitude matters more to action effectiveness than the complexity of an animal's milieu.

In this sense, the motionlessness of plants forces them to detect a multitude of stimuli imperceptible to animals. The fact that plants react appropriately to a very diversified environment gives them an extremely lush world. It is in fact much richer than the human world, which is limited to the perception of five senses (plants also detect and react to humidity, light, electromagnetism, etc.). The chemical messages that plants emit (more than 1,000 messages recorded to signal attacks) and the compounds they synthesize (which involve up to 200 molecules) are remarkably numerous and precise (Dicke and Bruin, 2001).

Uexküll adds that the probability of some mammal coming by directly underneath the spot where the tick is questing is so low that ticks must be perfectly adapted to their situation. They can thus remain motionless with no nourishment for a period of eighteen years if they perceive no stimulation which could cause them to react. But—in a way—what we perceive as a long period of time is only a moment for ticks. For Uexküll defines the moment as the smallest period during which one may not perceive any change. Thus, the experience of space and time is not absolute—rather it depends on one's milieu:

> Time, which frames all events, seemed to us to be the only objectively consistent factor, compared to the variegated changes of its contents, but now we see that the subject controls the time of its environment. While we said before, "There can be no living subject without time," now we shall have to say,

---

[30]   This is at least the case for Uexküll—but not for Heidegger (1992) who draws on it to claim that animals are poor in world.

> "Without a living subject, there can be no time" […] the same
> is true of space: Without a living subject, there can be neither
> space nor time (2010, p. 52).

The biological relation to time and space is not the same in all species. This difference is even greater in plants, since their sensitive sensors are both extremely numerous (*i.e.* the thousands of leaves and root tips) and extremely active (*i.e.* the thousands of different biochemical reactions induced by such perceptions). Such hypersensibility in plants does not necessarily entail a rapid depletion of their vitality–despite what the correlation of inactivity with longevity in ticks may suggest. In plants, even a world very rich in perceptions and actions can entail exceptional longevity, as in the case of trees. This means that it is clearly necessary to disconnect an organism's activity (their world of action) from locomotion or movement. Indeed, some electrical and chemical reactions in plants are crucial for their sensibility–although they are invisible. These reactions can also be very quick relative to an organism's often imperceptible capacity for motion.

Effect Space and Movement in Plants

According to Uexküll, the effect space is the space within which the body can move in six directions–up, down, front, back, right, left–thanks to a perpendicular reference system set by the internal ear. Relative to an animal's body, this reference frame is stable. One can orient oneself via motions which minimal unit is a directional step (a little less than one inch long in humans). Following Uexküll, the internal ear can regenerate a complete route by decomposing each of these motion units. It acts like a compass, helping animals find their way a posteriori without using any effectual perception (visual or other).

Beyond the zoocentric aspect in the use of the internal ear, for Uexküll, the space of action is connected to motion and locomotion. Consequently, can we argue for a space of action in plants? The answer is complex–because, unlike animals, motion in plants does not involve locomotion, and, more generally, action does not necessarily involve motion of a range perceptible by humans. It is indeed possible to claim that any reaction involves a motion, and that such motion is perceptible at some level. Thus, if we do not perceive most reactions in plants, it is because of the cellular and molecular levels involved, too small and too quick, or because of the macroscopic level of the whole organism, where growth occurs, which is too slow. However, when we say "motion", we usually mean some motion perceptible by humans (and not a mere change of state). Since our perceptive range is very far from plant activities, the way in which we consider plants is impacted. Incidentally, the discovery of photography and film led to a small revolution in our "philosophical" approach to plant life–by giving us access to motions which were imperceptible for our senses, even though they were known via experimentation and thus thought. Furthermore, the fact that we do not perceive plant motion in our own perceptual space does not entail that plants do not perceive their own minute motions in their own effect space.

Our understanding of plants' relation to their milieu depends crucially on what space and motion mean for plants. It has been scientifically established that plants possess a sense allowing them to distinguish up from down (gravitropism). By contrast, it is unclear whether plants can distinguish left from right or front from back due to their fixedness. For if plants are able to move to the left, to the right, or toward the front or back, they do not do so specifically–contrarily to the upward and downward motions they display. When a plant grows upward and orients its roots downward, it does so because of its perception of gravity. Upward and downward motions have an inherent meaning for plants: the search for light, water and nutrients. By contrast, if a plant grows leftward or backward, for instance, it is not because it perceives left or back–which would have a value in themselves–but because of a stimulus which it seeks or avoids (the stimulus being what the plant perceives).[31] In this case, the plant draws on its perception space rather than on its effect space to orient itself. Moreover, the body symmetry in a fixed plant organism differs from the one in a mobile animal. This could also account for differences between the effect space of plants and the one of animals as described by Uexküll. Like animals, plants are not disposed on either side of a horizontal symmetry axis, and their left and right sides are identical (bilateral symmetry). Unlike the bodies of vertebrate animals, the stems of plants are symmetrical in the back and front (*i.e.* morphologically, no distinction can be made in the front and back of a plant—while we can distinguish an animal's back from its belly). For these reasons, we call symmetry in plants radial. However, these symmetry axes are much more flexible and theoretical than in animals, since a plant's development remains mostly indeterminate and evolves following its growth—differently from an animal's development, which is strictly determined as early as embryogenesis.

Which organs perceive up and down? All developing organs in plants have specific cells called statocytes. They contain elements drawn by gravity (*i.e.* organelles called statoliths) which sediment at the bottom of the cell and thus orient the root's growth based on their perception of up and down. Finally, plants perceive their own body (*i.e.* proprioception). Consequently, they likely possess some minimal spatial coordination system (at least up/down) which defines their own effect space. However, due to their fixedness, effect space in plants cannot fulfill the role of compass Uexküll pointed out for it in animals—since plants do not move. It also remains to be seen whether a motion unit amounting to the directional step in animals exist in plants—and if it does, what unit it is. We could imagine this unit to be the minimal growth reaction

---

[31]   Like plants, animals give up and down a different value, but a mere contextual value to left and right. Furthermore, animals—unlike plants—give an inherent value to front and back since their morphology and sense organs imply locomotion in a preferred direction.

when plants react to a change in gravity or perceived lighting, or something similar.[32]

Space of Action and the Sense of Touch in Plants

For Uexküll, the effect world has a direct connection with the exploration the sense of touch allows. Touch defines the tactile space of an organism like so:

> In feeling out [an object], places connect themselves to directional steps, and both serve the process of image-formation (Uexküll, 2010, p. 61).

Their sensibility allows animals to orient their locomotion and to analyze objects in their environment. Such sensibility changes depending on the body part and species considered. Thus, in humans, the tongue and fingers are very precise while the back is not. Let us add that, as Aristotle had noticed, touch seems the most universal sense among living beings: it can be found in the most primitive or simplest species, like unicellular organisms or mollusks, which is not necessarily the case for the other senses. We can thus conceive that plants have a sense of touch, and therefore a tactile space.

Scientific experiments have shown that plants feel tactile stimuli to which they can react (involving cases of habituation and sensitization). They have allowed us to discover mechanisms of communication in plants. Thigmotropism is a movement elicited as a response to a contact and thigmomorphogenesis is a change in growth following tactile stimulation. In thigmotropism, movement is generally elicited by contact with a solid object—it allows roots to skirt around rocks and allows the twining stems of vines to twine around their support, or to attach to it in only a few hours thanks to their tendrils (Darwin, 1865; Chamovitz, 2013, chap. 3). Such movement originates in the decreasing growth rate of the cells on the lower side—in contact with the object—and

---

[32] For instance, do plants react to a one-degree change relative to their vertical axis? What about a change by half a degree? By one hundredth of a degree?

in the increasing growth rate of the cells situated on the outer side (Raven *et al.*, 2013, chap. 28).

All plants are sensitive to touch in a way which allows them to feel the wind, the attack of a predator, hot, cold or a rival plant–even though their sensibility does not elicit a movement directly perceptible by human senses. The result of this sensibility is most often a change in growth (*i.e.* thigmomorphogenesis). The *Sicyos angulatus* plants–twining plants from the cucurbits family–have a sense of touch nearly ten times more sensitive than humans: they can feel a fibril weighing only 0.0088 oz. The reactions of sensitive plants and Venus flytraps to the touch are perceptible to the human eye. While the tactile sensations in animals allow them to avoid pain, tactile sensations in plants allow them to adjust their development to adapt optimally to their environment (Chamovitz, 2013, p. 80–83), but do not a priori generate pain or any emotional reaction.

The tactile space as conceived by Uexküll must be qualified, because plant sensibility is proportionately more "passive" than it is exploratory. Since plants constantly monitor and evaluate countless variations in the environment in which they are fixed, they are sensitive to more tactile perceptions than animals who move in their environment in active exploration. But plants do possess exploratory abilities, since the majority of their activity is the exploration of soil through their innumerable root tips. This is also true for vines–in which tactile exploration is not limited to roots, but expands to include above-ground plant parts. Accelerated video recordings of growing vines speak for themselves[33]: the plant first makes circumnutation movements which grow larger in search for a support (exploratory movements); then, as soon

---

[33] <https://www.youtube.com/watch?v=9xMVKbU2O98 > (visited on January 27, 2020) and <https://www.youtube.com/watch?v=kpp6Vc43Qjk> (visited on January 27, 2020).

as there is a tactile contact with the support, an active sensibility to touch steps in to analyze it and twine around it.

Territoriality

According to Uexküll, in some animals, subjectivity is also expressed through territoriality, which inherently relates to exploratory movements (Uexküll, 2010, p. 102–107). For Uexküll, territory is strongly associated with locomotion. A territory implies that one can move about in it, like moles in their burrows and spiders in their webs. Despite the fixedness of plants, could their competition for light and soil resources involve some territoriality? Locomotion could turn out to be a non-discriminating element of territoriality–pertaining uniquely to the animal models selected by Uexküll. Some plants indeed seem to actively defend what we can think of as their territory by literally poisoning the soil with toxins secreted by their roots or by their leaves falling to the ground–this is called allelopathy. It prevents the germination and growth of rival plants in the area of influence where allelopathic plants exploit resources. The analogy with animals protecting their hunting territory does not seem out of place. Furthermore, such territoriality does not express only rivalry, even among plants. Indeed, plants can use the same means they use to spread toxins to attract insects which protect them against predators present on their territory (Turlings *et al.*, 1990; Beyaert *et al.*, 2012), or secrete substances in the soil which allow and arrange symbiosis with bacteria and mushrooms (Giovannetti *et al.*, 1996; Akiyama *et al.*, 2005; Oldroyd, 2013), thus contributing to the co-creation of a common territory.

Sense of Shape in Plants

For Uexküll (p. 79–85), the perception of shape and motion is seen only in superior animals because it implies to group different places together through some kind of centralization, which would be impossible, for instance, in sea urchins.

Besides, one cannot necessarily separate the perception of shape and the perception of motion in animals. Thus, jackdaws cannot recognize the motionless shape of their prey, grasshoppers–they can only perceive their shape and hunt them when they are in motion. More radically, for great scallops, only the speed of a motion counts (*i.e.* the speed of their predator's motion), while the shape and color of the thing in motion do not matter at all. Darwin thought that earthworms could distinguish shape, because they knew how they needed to grab leaves and pine needle bundles (by their tip and their base, respectively) in order to drag and fit them into their hole–as if they could recognize the shapes of the tip of leaves and the base of needle bunches. In fact, worms discriminate them based on the taste of the leaf tip or needle base: if we sprinkle pieces of leaf tip or needle base on small sticks whose base and tip are identical in shape, worms will grab them accordingly.

> So there was nothing to the notion of shape perception in earthworms. The question therefore became all the more urgent: Which animals could one assume to have shape as a perception sign in their milieus? (Uexküll, 2010, p. 84, transl. modified).

The sense of touch does not on its own imply a capacity to perceive shapes. But the result of the combination of the active movement of roots with the sense of touch allows us to suggest the hypothesis that plants could have a sense of shape.

When roots find stones on their way in the soil, they attempt to go around them–but if circumvention is impossible, a plant will stop directing other roots toward the obstacle. Similarly, if the obstacle can be circumvented, new roots will be directed to grow around it, rather than directly into it (Drénou, 2006). This would indicate that the obstacle's shape is integrated in a centralized manner. All vascular plants would then possess a sense of shape which simple animals do not have. It may appear unorthodox to argue for the possession of a sense of shape by vascular plants–but let us recall that, for Uexküll,

the sense of shape is connected with the exploratory faculties of an organism relative to his milieu. And while oysters
or jellyfish may not really need to explore their milieu, it is
certainly necessary for plants.[34] Being fixed, plants are forced
to optimize their knowledge of their environment through
exploratory motions.

Where above-ground plant parts are concerned, the way of
living influences the perception of shape. From a biological
perspective, vines truly perceive shapes. This is because
the "perception" of shape as shape is necessary for them to
select and adapt to an appropriate support in order to survive.
However, nothing here indicates that the "perception" of
shape could bring an adaptive advantage to above-ground
parts in other plants. Thus, it could be the same for Venus
flytraps as it is for earthworms or great scallops. If flytraps
capture prey, is it not because they react to stimuli of movement caused by a shape which they do not perceive in itself?
Besides movement stimuli, do flytraps not discriminate their
prey based on size? They could thus avoid snapping their
trap shut on prey too small or too large, which would force
them to reopen it shortly after–the way they do when they
have accidentally snapped on inedible matter. In fact, flytraps
do not close their traps on prey whose movement is insuffi-
cient to stimulate the reactive trigger hair, and conversely,
they do close their traps with no consideration for size if a
movement is sufficient, *i.e.* even if the captured prey is too
large to be digested. In this sense, flytraps and great scallops alike would perceive movement, but likely not shape,
since they do not perceive size. Likewise, in sensitive plants,
shape is not involved in reactions: the intensity of the contact
alone causes the leaves to fold inward reactively. The reaction can incidentally be triggered by the contact of something
with no shape, *i.e.* wind, fire or electricity. From such a

---

[34] A jellyfish is a plankton organism, *i.e.* its movement is not active, but results
from drifting in the current.

perspective, there would be some plants – perhaps a minority of them – whose above-ground parts perceive shape (among which vines) and others in which that is not the case. And if some plants perceive shape in their milieu, it is due to its signification relative to their way of living.

But saying that perception of shape is exceptional in plants based on its infrequency in above-ground plant parts would be to forget that all vascular plants have a sense of touch through their roots and the capacity to identify and circumvent obstacles in the soil through growing motions (the latter more or less equivalent to directional steps in animals). Thus all plants would perceive shape, in Uexküll's sense, at the underground level. In addition, since the information originating in the roots are transmitted and distributed to the whole plant, the idea that the sense of shape in plants could be found in a precise location may require qualification.

The Visual Space of Plants

Do plants have a sense of sight allowing them to analyze the space around them? When discussing visual space, Uexküll takes in consideration animals who have no eyes – and nonetheless identifies photosensitivity in them. Since it focuses less on vertebrates, such "sight" allows more easily for the inclusion of plants in Uexküll's analysis of sight and visual space, as the following shows:

> Eyeless animals that, like the tick, have skin that is sensitive to light will most likely possess the same skin areas for the production of local signs for light stimuli as well as for tactile stimuli. Visual and tactile places coincide in their milieus.
>
> Only with animals that have eyes do visual and tactile places clearly separate (Uexküll, 2010, p. 61, transl. modified).

Even if plants have no eye or skin, their leaf surface is sensitive to the touch and to light.

Uexküll is interested in the way the eye generates a perceptually complex space by allowing for the more precise stimuli localization. The sense of sight creates a horizon which sets the farthest distance within a visual space (Uexküll speaks of a bubble surrounding animals). In eyeless organisms, there is no farthest distance and no horizon.[35] What does this imply for plants? Can their photosensitivity give them access to a visual space distinct from their tactile space?

Chamovitz (2013, chap. 1) explains that plants are able to detect light and darkness as well as the intensity, origin and color of light. They thus detect the shadow of an object, or a rival plant above them. He maintains that plants possess a true sense of sight, citing the Merriam-Webster's definition of sight:

> The physical sense by which light stimuli received by the eye are interpreted by the brain and constructed into a representation of the position, shape, brightness and usually color of objects in space (Chamovitz, 2013).

But, from the outset, this definition opens a rift between sight and photosensitivity in plants. Even though sight and photosensitivity share some properties, they differ in many respects. It is thus true that both involve a perception which involves the physiological sense through which light stimuli are received by photoreceptors (in the retina or plant tissue). However, the mechanism and the finality of the two senses differ. For plants have no organ of sight, and no brain to interpret visual data. Consequently, the finality of such mechanism cannot be a "representation of the position, shape, brightness and color of objects in space". Both cases involve a perception of these data (location, shape, brightness, color), but it is not mediated by a representation: it is instantly transformed in a signal. Photosensitivity in plants is a kind of perception which differs in nature from eyesight.

Indeed, light turns out to be more than a mere signal for plants, since it is a direct "food source" or a source of energy allowing nutrients to be metabolized through photosynthesis.

---

[35] This is not entirely true to the extent that some eyeless animals can have a representation of space, distance and some kind of "horizon" via echolocation.

Plant leaves have numerous photoreceptors distinguishing between a wide array of colors which are not all visible to the human eye–for instance far-red (situated at a wavelength between 710 nm and 780 nm just before infrared) and ultraviolet.

Due to their sensibility to different wavelengths in the visible spectrum, plants distinguish shade as well as the quality of the light they receive. This enables them to perceive what projects shade above them and what catches some part of their light, and they adapt their growth accordingly. A plant can thus distinguish another plant from some odd obstacle, because the light reflected by its foliage differs in nature from the one reflected by a rock (Sultan, 2015, p. 60–62). By analyzing light, plants could similarly recognize kin (Crepy and Casal, 2015). In this sense, plants' photosensitivity plays a direct role in their relation to space (as is the case in animals). By moving naturally toward a source of more intense light, plants in fact move toward an unobstructed space. Conversely, when plants avoid the shade, they move away from a space occupied by an obstacle or rival which prevents the spatial exploitation of their milieu and its resources. Finally, the spatial milieu of plants is probably not a represented visual totality–it is rather a network containing points of interest or affordances. However, some papers hypothesize that some plants could discriminate shapes and colors of objects through eye-like ocelloids[36] (Gavelis *et al.*, 2015; Baluska and Mancuso 2016).

Lastly, plants also perceive the intensity of light–they adapt to it by changing the orientation of their leaves, trying either to increase the amount of light received by maximizing the exposed surface, or oppositely to decrease it to avoid burns.

---

[36] An ocelloid is a complex subcellular organ found in the unicellular family of dinoflagellates. Its structure and its function are similar to a multicellular eye in the way it detects light thanks to an iris, cornea, lens, and retina.

Plants, thanks to their photosensitivity, are therefore able to obtain information about location, shape, brightness and color much like eyes do. But the analogy is functional and works only partially. In addition, it should not conceal the fact that plants' sensibility to light greatly exceeds the function of an eye. Indeed, light reception and discrimination are probably infinitely more crucial for the life of plants than for the human eye whose prime role is to orient us in space (and the same is likely the case for many animals). While a human being can live a fulfilling life with color-blindness, achromatopsia or even total blindness, the perception of light wavelengths is vital to plants. It regulates the growth of stems and leaves, germination, the orientation of chloroplasts, photosynthesis, flowering, phototropism, photoperiodism, pigment synthesis, etc. Finally, in plants, sensibility to light is not used so much to conceive space, as it is used in animals, but rather to conceive time.

Relation to Time and Plant Temporality

Photoperiodism strongly influences plant lifecycles (growth, reproduction, dormancy). Photoperiodism measures the length of days: for instance, it allows plants to anticipate nightfall or to "know" when to flower over the course of the year. So-called short-day plants flower when days become shorter than a determined length, while "long-day" plants flower when days become longer than a determined length. In fact, plants do not measure the length of the days, but rather the periods of uninterrupted obscurity. Short-day plants are then plants sensitive to long nights and long-day plants, plants sensitive to short nights. By exposing them to light overnight, even for a few minutes, the former stop flowering and the latter begin flowering. Just like Darwin had shown that blue light guided phototropism thanks to buds, scientists studying photoperiodism have shown that plants relied only on red light to estimate the length of nights. Being short-day plants, irises will flower if exposed to red light during a short period at nighttime.

Such perception of light periods is closely connected to memory (even though it is not based on a conscious, reflexive act of recollection), and, consequently, to temporality. The capacity of plants to discriminate light color, but the intensity and origin of light as well, also plays a role in their relation to time.

For instance, cryptochromes – a kind of photoreceptor found in plants and animals – controls the regulation of circadian rhythms and already existed in unicellular organisms from which plants and animals have evolved following divergent paths. Consequently, there is a homology between plant photosensitivity and sight in animals and humans both in the structure of their mechanism (cryptochromes) and in the finality of light processing (the control of circadian rhythms). Thus, our eyes also allow us to measure the passing of time. Based on this, one can hypothesize that the clear distinction between time perception and space perception could be the result of an evolutionary diversification in organisms originating in a common starting point where spatiotemporal perception would have been less differentiated. Among animals, the sense of sight is now primarily used for spatialization, but also, to a certain extent, for temporalization (through circadian rhythms). Meanwhile, in plants, light likely plays a temporal role as crucial as its spatial role. However, temporality in plants is not as linear as it is in animals – it is rather rhythmical, following the variations of their environment.

Plant temporality can be called rhythmical by contrast with a linear temporality because the metabolism of plants closely depends on conditions in the outside environment, like sunshine and temperature. Trees losing their leaves in the winter thus slow down their metabolism, which contracts the time they experience; likewise, they experience a more dilated time in the spring. Changes in lived temporality are not only related to temperature in plants: drought or dearth can have the same effect. Plants can slow down their life rhythm (and sometimes even stop it) to avoid death and wait for better days. By contrast, human being – defined by a mostly linear experience of time – cannot adapt their metabolism in the same way: they are forced to undergo the passing time uniformly (from a biological rather than a psychological point of view), even in the event of freezing cold or famine which would quickly precipitate their death.

A specific fundamental difference in the experience of lived time is likely found in whether one's metabolism is homeothermic ("warm-blooded") or not. A constant temperature maintained at all times requires a linear metabolism. Poikilotherm animals, in contrast, adapt their metabolism to the outside environment – which implies that their temporality is not linear, but rather rhythmical. The boundary between rhythmical time and linear time is not necessarily clear-cut, and both temporalities coexist to varying degrees in most living beings, for instance in the cycle of sleep or hibernation. Incidentally, all living beings depend in part on a rhythmical temporality through their circadian rhythms.

Lived experience differs in plants and animals based on their respective relation to time. At each moment of our life, we can picture the way in which our experience unfolds relative to the life expectancy and maximal longevity of our species. Even if animals can have no representation of the total amount of time they potentially have left to live, they undergo some biological effects as they reach old age, which unavoidably impacts their lived experience. Conversely, perennial plants such as trees do not grow old strictly speaking after they have reached sexual maturity. Trees do not undergo programmed senescence like animals do. In this sense, they are sometimes said to be potentially immortal. Death is not internally programmed as a deadline in the metabolism of trees – it comes from an outside source (Lenne, 2014, p. 25). As they are able to regenerate, trees never die from old age, since they do not age–they always die in their prime from accidental causes. The experience of perennial plants therefore does not depend on a relation to finitude and its effects, but on the conditions of an optimal metabolism.

Perception and Reactions in Plants

According to Uexküll, perception time is relative to the time units perceptible by a given species–moments. Humans can thus perceive time units as short as one eighteenth of a second. This is why a film must have at least eighteen images per second, unless it be perceived as jagged. However, a perceptible moment can only ever be deduced. For instance, one stimulates an animal at different speeds and observes when it reacts and no longer reacts. But since what one measures is more an observed reaction than the perception of the moment itself, it is a behavior rather than a perception which is being evaluated. It is therefore possible that an animal may perceive a stimulus below or beyond its reactive range.

In fact, this is the case in animals–but also in plants: in some cases, stimuli accumulate and lead to a reaction only when a certain threshold is met. These stimuli are thus perceived without eliciting a (perceptible) reaction at each perception. However, Uexküll concludes that when animals act, it is because they perceive, and if they do not act, it is because they do not perceive. Yet in experiments conducted on plants, there may be cases of delayed reactions to stimulation. For instance, if a Venus flytrap's trigger hair records a first excitation, the trap does not close–the plant does not react. To close, it waits for a second stimulation–which must occur within a few seconds–while keeping the first stimulation in mind (which prevents it from shutting its trap on some non-living thing). Thus, plants can "perceive" while displaying no behavioral reaction perceptible by humans.[37] But according to Uexküll, reaction is the "psychic" manifestation of a behavior – it can thus not be assimilated to a purely mechanical, arduously perceptible biochemical reaction. Yet the case of stimulations accumulated and preserved until a delayed reaction occurs signals that the psychic and physiological aspects of behavior may be mutually irreducible – even in plants.

Effect Tone

According to Uexküll, knowledge of a milieu involves a combination of perceptions and actions. Thus, a perception-image is most often connected to an action-image. The action-image is the representation we may have of the action which a perception implies we could perform. For instance, in humans, seeing an armchair implies the action "sitting", seeing a cup, the action "drinking". In their milieu, animals generally only perceive these objects which have effect tone (or affordance), which explains how their actions come to have a high degree of certitude. A milieu containing fewer objects also means less choice for an organism (Uexküll, 2010, p. 73–75). For instance, since paramecia (unicellular eukaryotes) merely run from all they touch, if they possessed an action-image of their activities, their milieu would be composed of similar objects, all of them having the connotation "obstacle". Such a milieu would be very high in certitude. One must speak in the conditional here – for Uexküll's classification based on action-images implicitly emphasizes the role of representations in object perception. But taking this terminology too literally may keep one from understanding plant behavior in a non-representational way.

---

[37] The end of dormancy in tree buds is a more general example of a reaction caused by the accumulation of stimuli – in this case, cold. Some varieties of apricot trees thus require 650 hours of mean daily temperature below 45°F to begin budding while some apple trees require 1,500 hours.

Indeed, even if plants, like paramecia, have no representations (or, at least, not as we have), a plant's milieu is likely richer than a unicellular's, because the plant can detect a multitude of different stimuli and react to them accordingly by performing different actions. Paraphrasing Uexküll, one could say that the plant's world would not be inhabited by a multitude of copies of one type of object. Light implies for plants to grow toward it, dampness implies for them to direct their roots in some direction, the contact of the wind, to strengthen their stem, an attack by parasites, to produce toxins, etc. Plants can even react to stimuli that appear extremely similar to us by performing different actions. Thus a young tree finding itself in the shade will modify the way it grows (*i.e.* by avoiding shade) depending on the shade's origin – whether it emanates from another tree (a rival), or from an object, like a stone (which is not a rival). The same goes for roots, since – as we have previously seen – plants react differently depending on which obstacle their roots meet: an inorganic obstacle, their own roots, the roots of kindred plants or those of rival plants. Piercing a leaf with a needle will trigger a different reaction from an identical injury caused by an insect, because plants are able to detect compounds from their predators' saliva and to react accordingly (Zebelo and Maffei, 2012). The world of plants is therefore much less certain and determined than the world of paramecia (at least as Uexküll conceived it), because many more possibilities (or even choices) are open to them.

Conclusions on Plant Ethology: Subjectivity and Milieu

Even if Uexküll explicitly confines his ethology to animals and their milieu, many of his ideas can be adapted to suit a certain conception of plants. There may be little structural homologies between animals and plants (save for circadian rhythms), but the finalities of perception seem to overlap at least in part in the way organisms give meaning to their milieu. It is precisely this latter aspect which turned out to be philosophically crucial in our inquiry into biosemiotics, since a behavioral act always depends on a finality (the simplest finality being a need).

 Uexküll evokes a plant only once, at the end of his book – an oak tree. He first takes the oak, not as a subject with its milieu, but as a milieu for other organisms: foresters, but also foxes who find shelter between its roots, or owls, among its branches, and "many animal subjects" (2010, p. 126) including a multitude of insects. In fact, each part of the oak tree is a part of the milieu of a given animal and provides it an effect tone (the branches support the nest, the hollow between the roots serves as a shelter, etc.). But does associating plants with an animal milieu mean that the oak tree cannot also be a subject of its own?

> In the hundred different environments of its inhabitants, the oak plays an ever-changing role as object, sometimes with some parts, sometimes with others. The same parts are alternately large and small. Its wood is both hard and soft; it serves for attack or for defense. If one wanted to summarize all the different characteristics shown by the oak as an object, this would only give rise to chaos. Yet these are only parts of a subject that is solidly put together in itself, which carries and shelters all milieus – one which is never known by all the subjects of these milieus and never knowable for them. (Uexküll, 2010, p. 132, transl. modified).

It would thus be possible to consider plants as subjects, even if they appeared to us and to other animals like a milieu above all. In the dualistic framework of modern naturalism, plants are solely objects – they are part of the landscape. But in Uexküll's text, the oak tree, like the rest of living beings, is an object and also a subject. Uexküll explicitly calls it a "subject solidly put together in itself". Are we then talking about a strictly organic subject? Such organic subjectivity – which includes plants – is not mechanistic. As Canguilhem says, what distinguishes organisms from machines "is precisely that its purpose, in the form of its totality, is present to it and to all its parts" (Canguilhem, 2012, p. 76). Subjectivity in this weaker sense would serve to orient finalities and guide the fulfillment of needs in any organism (by contrast with a programmed machine which does not adapt and does not reproduce). Finality as a source for the unification of an organism's behavior can be likened to Uexküll's subjectivity, and its manifestations can appear as a result of the study of a cohesive behavioral unit. In this sense, plants could have a unifying subjectivity despite their lack of morphological individuality and their blurred functional limits.

But beyond being subjects, plants are indeed milieus since they are conditions of possibility for the life of almost all other organisms depending on them in some way (including humans). Incidentally, their labile individuality shows their flexible compatibility with the rest of their milieu. It would then be preferable to think of plants above all through their processes and their constitutive roles in milieus, rather than merely as individuals or subjects.

Conclusion

A general study of behavior tells us that the lines between the mineral, plant and animal kingdoms are blurred—by contrast with the historically popular belief concerning their separation and hierarchy (in which humans would happen to occupy a superior position). We now know that plants are sensitive and can react in various ways. Movement is only one of the possible manifestations of these behaviors. It is part of a larger process including sensibility, information processing and perceptible reactions. In plants, reactions are often connected to growth or changes in internal states.

According to recent experiments, all living beings, including plants, appear to possess faculties of communication, memory, learning and even, following some authors, consciousness or intelligence. Yet this does not justify anthropomorphism or the relativism which results from a lack of differentiation between kinds of beings and organisms. The aptitudes plants display are not found in inorganic things, and they often appear in degrees or modalities different from those present in "superior" animals and humans. It is then abusive to liken plant behavior to the passive or strictly mechanical behavior of physical objects. All life forms possess elementary cognitive aptitudes which are justified in the light of evolutionary biology. Some objectives common to all living beings may have been selected and favored through similar aptitudes, or even similar structures. Some functions are found in plants which converge with those of a brain–like non-nervous sensibility, memory, or learning at the cellular level. But analogous behavioral aptitudes do not mean that plant behavior is identical to animal behavior. A plant-based perspective generates other ways of thinking about concepts, divergences with the modern naturalist tradition, and criticisms – yes – but it also opens up new possibilities. The mind-body dualism, the claim that only some species could possess cognition or the claim that some faculties (like memory and learning) necessarily imply interiority and mental representations can thus be questioned anew, and lead to the development of alternative accounts of behavior.

For this reason, we should be open-minded with cultures more inclusive of plants. Not so that we could naively transpose their ways of thinking onto our scientific method – but so that we could relativize our own conceptions or preconceptions, and perhaps even question the scientific practices they led to.

However, as biologists, specialists of plant behavior remain above all mindful of the rigorous scientific methods of their community. The practice of contemporary Western science is not one with the cultural practices of Amazonian Indigenous peoples, the naturalist paradigm is not one with animism. However, an author such as Matthew Hall, in *Plants as Persons* (2011), leads an explicit dialog between non-Western cultures and plant neurobiology in order to put forward a more unified ethical conception, not limited to the sole adaptive biological or cultural approach – while including experimental behavioral science. Even though, from the naturalist point of view, plants have no intentions and no reflexive consciousness, scientific results alone must not solve all problems on their own, especially where ethics is concerned. Indeed, our relations with other living beings–including plants–are not limited, or should not be limited, to a strictly naturalist, techno-scientific relation.

Controversies about the study of plant behavior most often emerge from a reductionist epistemological framework which, incidentally, has likely impeded the acknowledgement of some of their aptitudes due to artificial laboratory conditions (Thellier, 2015, p. 73). Laboratory experiments must not either be systematically opposed to experimentation and observation performed in natural settings, as if the latter were unproblematic. The history of botany and the more recent history of behaviorism show that theoretical frameworks from a given time and discipline determine what discoveries are possible, but also the very formulation of the problems and questions we pose. Theoretical concepts and choices are never neutral and questions are never innocent. They can lead to new hypotheses and new results. Just as the modern mechanistic framework prevented the acknowledgement of true sensibility in plants, a too reductionist or too artificial physiological framework may have prevented the acknowledgement of some behaviors which as a result of such lack of acknowledgement, remained unknown.

Moreover, it can be difficult to detach a strictly biological (or vegetative) behavior from a psychic or cognitive (animal) behavioral level, since even the "simplest" organisms display memory and learning. More broadly, the sensibility to one's environment and the aptitude to survive presuppose a network of elementary significations based on finalities and values in any living being. In plants, what is valuable is minimally what is recognized as self or non-self, or what constitutes a threat or a useful resource in the environment. These behaviors involve the organism in the resolution of the problems it encounters. They can be interpreted by some authors as forms of consciousness or intelligence – if these faculties are understood in a redesigned, non-anthropomorphic sense. Recent trends in biology – like behavioral ecology and niche constructions – converge toward these interpretations by pondering a better integration of organisms in their milieus (Hiernaux, 2018b). The study of plant behavior considered in a more systematic way – closely connected to a plant's environment – can lead to an externalized perspective on cognition and mind, *i.e.* cognition and mind conceived as relational properties, rather than as internal capacities (Rowlands, 2010). Plant behavior thus brings us to reconsider relations between living beings and their milieu and, ultimately, theories of human knowledge. Let us then be open-minded regarding these possible conceptual developments. Not only do they make possible new scientific experiments and potential discoveries – they also, more broadly, impact our philosophical understanding of plants, animals, humans and their relation to the world.

It remains possible, however, to single out the psychic level of behavior, by giving a much stricter definition of cognition – *i.e.* one which supposes mental representation or conscious intentions. This is a fundamental choice. Are we ready to recognize in each organism some form of autonomy and openness of its own – or do we intend to restrict such privilege to humans and so-called superior animals? The question cannot be answered solely on scientific grounds. What do we have to win or to lose by granting plants autonomy? On the one hand, to defend plant cognition amounts to letting go of the burdening traditional metaphysical dualisms of mind and body, and subject and object, where plants are passive, unintelligent and without much value. On the other hand, this involves renouncing methodological reductionism as an all-encompassing principle and setting the study of behavior in motion under the guidance of some form of methodological pluralism (Cvrcková *et al.*, 2009; Cvrcková *et al.*, 2016;

Hiernaux, 2019, 2021). The controversies will remain, but they are what drives scientific activity. If enactivism or plant neurobiology improve our knowledge of plant behavior, scientific goals are reached. The stakes are also pragmatic in nature–given the option of an unchanging and historically dated framework, with its accompanying close-mindedness, and the option of a richer framework allowing us to make novel hypotheses, and to invent new apparatus (to test new learning or sensibilization modalities in plants, for instance), do we not have a heuristic, or ethical interest to opt for the latter and take the risk of thinking differently?

Ethics should not ignore the scientific reality of the entities it discusses. Ethology taught us to better understand and consider animals. The study of plant behavior should do the same. A scientific and philosophical analysis of plant behavior also teaches us that it differs in many respects from animal behaviors involving emotions and suffering. This is an important point. It allows us to reject both the claims of anthropomorphic plant ethics and the one brought forward by those maintaining that the moral treatment of plants would negatively impact animal well-being. For to simply transpose pathocentrist or utilitarian ethics onto plants does not seem judicious, since plants experience a priori no emotions or suffering and cannot be individualized the way animals can. Likewise, to purely exclude plants from the moral community (sometimes for the so-called benefit of animals) is just as problematic given our current ecological knowledge and its implications. The sensibility and capacities of plants are well suited for their fixed, autotroph relation with their environment, which ecology and evolution can incidentally account for. The study of plant behavior plays a part in the realization that ethics of life should incorporate organisms and their relations with their milieu, and therefore become a part of environmental ethics. For plant behaviors show particularly clearly that the activities of living beings–especially autotroph organisms–have a crucial ecological impact, and are then important constituents of milieus. Yet plants cannot be reduced to passive matter or abstract environment. Living, sensing, and able to learn and to solve problems, they persevere in their being and can also play an important role in the human and non-human community (at a cultural, agronomic and ecological level). This should entail some respect and greater open-mindedness toward legal reflections (which incidentally already exist in various forms).

Attributing to plants a behavior analogous to our own—yet very different—and attempting to understand its specificities using a positive inquiry rather than an approach based on the devaluation or exclusion of plants with respect to animals—as I have done in this book—can thus result in more consideration for plant life. To admit that a plant is sensitive to its environment, its own interests and vital purposes, even without being conscious of it, could help extend to them various forms of legal and moral respect, without going as far as claiming that plants are sacred or feel suffering (Hallé, 1999; Hall, 2011; Marder, 2013b). The declaration by the Swiss Federal Ethics Committee on Non-Human Biotechnology (ECNH, 2008) specifically takes into account the moral consideration of plants with regards to their own interest (Pouteau, 2014, 2018). To admit that plant sensibility exists entails no plea for the plants-have-feelings-too argument granting feelings to plants at the expense of animals, nor does it require the rejection of distinctions between orders or species. What the acknowledgement of plant sensibility and the study of associated behaviors can teach us is to better understand and experience the diversity of that which connects living beings—life.

References

Afeissa H.-S., Jeangène Vilmer J.-B. (eds.), 2010. *Philosophie animale : différence, responsabilité et communauté*, Paris, Vrin.

Akiyama K., Matsuzaki K.-I., Hayashi H., 2005. Plant sesquiterpenes induce hyphal branching in arbuscular mycorrhizal fungi. *Nature,* 435(7043), 824-827.

Alpi A., Amrhein N., Bertl A., Blatt M.R., Blumwald E., Cervone F., *et al.*, 2007. Plant neurobiology: no brain, no gain?. *Trends in Plant Science*, 12, 135-136.

Arber A., 2012. *The natural Philosophy of plant form*, Cambridge, MA, Cambridge University Press (1st ed. 1950).

Aristotle, 1961. De anima: William David Ross (ed.), Aristotle. *De anima*, with introduction and commentary. Oxford, Clarendon Press, 1961.

Atlan H., 1999. *La fin du tout génétique. Vers de nouveaux paradigmes en biologie*, Versailles, Éditions Quæ (coll. Sciences en questions).

Babikova Z., Gilbert L., Bruce T.J.A., Birkett M., Caulfield J.C., Woodcock C., Pickett J.A., Johnson D., 2013. Underground signals carried through common mycelial networks warn neighbouring plants of aphid attack. *Ecology letters*, 16, 835-843.

Backster C., 2014. *L'intelligence émotionnelle des plantes*, Paris, Trédaniel (1st ed. 1966).

Baldwin I.T., Halitschke R., Paschold A., von Dahl C.C., Preston C.A., 2006. Volatile signaling in plant-plant interactions: 'talking trees' in the genomics era. *Science*, 311(5762), 812–815.

Baldwin I.T., Schultz J.C., 1983. Rapid changes in tree leaf chemistry induced by damage: evidence for communication between plants. *Science*, 221, 227-279.

Baluška F., Mancuso S., Volkmann D. (eds), 2006. *Communication in plants: neuronal aspects of plant life*, New York-Berlin, Springer.

Baluska, F., Mancuso S., 2016. Vision in plants via plant-specific ocelli? *Trends in Plant Science*, 21, 727-730.

Barlow P.W., 2008. Reflections on 'plant neurobiology'. *BioSystems*, 92(2), 132-147.

Bastien R., Bohr T., Moulia B., Douady S., 2013. Unifying model of shoot gravitropism reveals proprioception as a central feature of posture control in plants. *Proceedings of the National Academy of Sciences of the United States of America*, 110(2), 755-760.

Bernier G., 2013. *Darwin un pionnier de la physiologie végétale. L'apport de son fils Francis*, Bruxelles, Académie royale de Belgique.

Beyaert I., Köpke D., Stiller J., Hammerbacher A., Yoneya K., Schmidt A., Gershenzon J., Hilker M., 2012. Can insect egg deposition "warn" a plant of future feeding damage by herbivorous larvae ?. *Proceedings of the Royal Society B*, 279(1726), 101-108.

Boscowitz A., 1867. *L'âme de la plante*, Paris, Ducrocq.

Bourgeade P., Boyer N., De Jaegher G., Gaspar T., 1989. Carry-over of thigmomorphogenetic characteristics in calli derived from *Bryonia dioica* internodes. *Plant Cell, Tissue and Organ Culture*, 19, 199-211.

Bournérias M., Bock C., 2006. *Le génie des végétaux. Des conquérants fragiles*, Paris, Belin.

Braam J., Davis R.W., 1990. Rain-induced, wind-induced, and touch-induced expression of calmodulin and calmodulin related genes in Arabidopsis. *Cell*, 60, 357-364.

Brenner E.D., Stahlberg R., Mancuso S., Baluška F., van Volkenburgh E., 2007. Response to Alpi *et al.*: Plant neurobiology: The gain is more than the name. *Trends in Plant Science*, 12(7), 285-286.

Brenner E.D., Stahlberg R., Mancuso S., Vivanco J., Baluška F. & Volkenburg E., 2006. Plant neurobiology: An integrated view of plant signalling. *Trends in Plant Science*s, 11, 413-449.

Burgat F. (ed.), 2010. *Penser le comportement animal*, Versailles, éditions Quæ.

Burgat F., 2006. *Liberté et inquiétude de la vie animale*, Paris, Kimé.

Cahill J.F., 2015. Introduction to the special issue: Beyond traits: Integrating behaviour into plant ecology and biology. *AoB Plants*, 7: plv120.

Calvo P., Baluška F., 2015. Conditions for minimal intelligence accross Eukaryota: a cognitive science perspective. *Frontiers in Psychology*, 6, 1329.

Calvo P., Trewavas A., 2020. Physiology and the (Neuro)biology of Plant Behavior: A Farewell to Arms. *Trends in Plant Science*, doi:10.1016/j.tplants.2019.12.016.

Calvo P., 2016. The Philosophy of Plant Neurobiology: a manifesto. *Synthese*, 193(5), 1323-1343.

Canguilhem G., 1965. *La formation du concept de réflexe aux XVIIe et XVIIIe siecles*, Paris, PUF.

Canguilhem G., 2008. *Knowledge of Life* (transl. Stefanos Geroulanos and Daniela Ginsburg), New York, Fordham University Press.

Canguilhem G., 2012. "The Problem of Regulation in the Organism and in Society" in *Writings in Medicine* (transl. and ed. Stefanos Geroulanos and Todd Meyers), New York, Fordham University Press, 67-78.

Cesalpino A., 1583. *De plantis libri*, Florentia, Georgium Marescottum.

Chamovitz D., 2013. *What a Plant Knows: A Field Guide to the Senses*, New York, Scientific American/Farrar, Strauss & Giroux.

Churchland P.S., 1986. *Neurophilosophy: Toward a unified science of the mind-brain*, Cambridge, MA, MIT Press.

Churchland P.S., 2002. *Brain-wise: Studies in neurophilosophy*, Cambridge, MA, MIT Press.

Clarke E., 2010. The problem of biological individuality. *Biological Theory*, 5(4), 312-325.

Clarke E., 2012. Plant individuality: a solution to the demographer's dilemma. *Biology and Philosophy*, 27, 321-361.

Coccia E., 2016. *La vie des plantes, une métaphysique du mélange*, Paris, Rivages.

Corbin A., 2014. *La douceur de l'ombre. L'arbre source d'émotions, de l'Antiquité à nos jours*, Paris, Flammarion.

Crepy M.A., Casal J.J., 2015. Photoreceptor mediated kin recognition in plants. *New Phytologist*, 205, 329-338.

Cvrcková F., Lipavská H., Žárský V., 2009. Plant intelligence: why, why not or where? *Plant Signaling & Behavior*, 4, 394-399.

Cvrcková F., Žárský V., Markoš A., 2016. Plant Studies May Lead Us to Rethink the Concept of Behavior. *Frontiers in Psychology*, 7, 622.

Darwin C., Darwin F., 1880. *The power of movement in plants*, London, John Murray.

Darwin C., 1865. On the movements and habits of climbing plants. *Journal of the Linnean Society of London* (Botany), 9, 1-118.

Delaporte F., 2011. *Le second règne de la nature*, Paris, Éditions des archives contemporaines (1st edition 1979), (Engl. transl. Arthur Goldhammer, *Nature's Second Kingdom: Explorations of Vegetality in the Eighteenth Century*, Cambridge, MA, and London: MIT Press, 1982).

Delattre A., 2019. *Le koudou et l'acacia : histoire et analyse critique d'une anecdote*. Tela Botanica, <https://www.tela-botanica.org/2019/03/le-koudou-et-lacacia-histoire-et-analyse-critique-dune-anecdote/> (visited on January 27, 2020).

Dennett D.C., 1983. Intentional systems in cognitive ethology: The 'panglossian paradigm' defended. *The behavioral and brain Sciences*, 6, 343-390.

Descola P., 2005. *Par-delà nature et culture*. Paris, Gallimard.

Despret V., 2002. *Quand le loup habitera avec l'agneau*, Paris, Seuil.

Despret V., 2009. *Penser comme un rat,* Versailles, éditions Quæ.

Despret V., 2014. *Que nous diraient les animaux si... on leur posait les bonnes questions ?,* Paris, La Découverte.

Dicke M., Bruin J., 2001. Chemical information transfer between plants: Back to the future. *Biochemical Systematics and Ecology*, 29, 981-994.

Drénou C., 2006. *Les racines : face cachée des arbres*. Paris, Institut pour le développement forestier.

Dretske F., 1988. *Explaining Behavior,* Cambridge, London, MIT Press.

Drouin J.-M., 2008. *L'herbier des philosophes*, Paris, Seuil.

Dumais J., 2013. Beyond the sine law of plant gravitropism. *Proceedings of the National Academy of Sciences of the United States of America*, 110(2), 391-392.

Errera L., 1910. *Recueil d'œuvres. Physiologie générale, Philosophie,* Bruxelles, Lamertin.

Federal Ethics Committee on Non-Human Biotechnology (ECNH), 2008. The dignity of living beings with regards to plants. Moral consideration of plants for their own sake. En ligne, <https://www.ekah. admin.ch/inhalte/_migrated/content_uploads/ e-Broschure-Wurde-Pflanze-2008.pdf> (visited on January 27, 2020).

Firn R., 2004. Plant intelligence: an alternative point of view. *Annals of Botany*, 93, 345-351.

Fox Keller E., 1999. *La passion du vivant. La vie et l'œuvre de Barbara McClintock/prix Nobel de médecine,* Paris, Sanofi-Synthélabo.

Gagliano M., 2015. In a green frame of mind: perspectives on the behavioural ecology and cognitive nature of plants. *AoB PLANTS*, 7, plu075; doi:10.1093/aobpla/plu075 .

Gagliano M., Renton M., Depczynski M., Mancuso S., 2014. Experience teaches plants to learn faster and forget slower in environments where it matters. *Oecologia*, 175(1), 63-72.

Gagliano M., Vyazovskiy V.V., Borbély A.A., Grimonprez M., Depczynski M., 2016. Learning by association in plants. *Scientific Reports,* 6, 38427.

Garbaye J., 2013. *La symbiose mycorhizienne.* Versailles, éditions Quæ.

Gavelis G. S. *et al.*, 2015. Eye-like ocelloids are built from different endosymbiotically acquired components. *Nature*, 523, 204-207.

Gerber S., 2018. An herbiary of plant individuality. *PTPBio. Philosophy, Theory, and Practice in Biology*, 10(5), 1-5, doi:10.3998/ptpbio.16039257.0010.005.

Gibson J. J., 1979. *The Ecological Approach to Visual Perception*, Boston, Houghton Mifflin.

Giovannetti M., Sbrana C., Citernesi A., Avio L., 1996. Analysis of factors involved in fungal recognition responses to host-derived signals by arbuscular mycorrhizal fungi. *New phytologist*, 133(1), 65-71.

Godfrey-Smith P., 2016. Individuation, subjectivity and minimal cognition. *Biology and Philosophy*, 31(6), 775-796.

Greene E.L., 1909. *Landmarks of Botanical History I.* Washington, Smithsonian Institution.

Gruntman M., Novoplansky A., 2004. Physiologically mediated self/non-self discrimination in roots. *Proceedings of the National Academy of Sciences of the United States of America*, 101, 3863-3867.

Guerra S., Peressotti A., Peressotti F., *et al.* 2019. Flexible control of movement in plants. *Scientific Reports*, 9, 16570.

Hall M., 2011. *Plants as Persons: a Philosophical Botany*, Albany, NY, State University of New York Press.

Hallé F., 1999. *Éloge de la plante : pour une nouvelle biologie*, Paris, Seuil.

Hegel G.W.F., 2004. *Philosophie de la nature*. Trad. fr. Bourgeois B., Paris, Vrin (1st ed. 1817).

Heidegger M., 1992. *Les concepts fondamentaux de la métaphysique (1929-1930)*, Paris, Gallimard.

Heil M., Silva Bueno J.C., 2007. Within-plant signalling by volatiles leads to induction and priming of an indirect plant defense in nature. *Proceedings of the National Academy of Sciences of the United States of America*, 104, 5467-5472.

Hiernaux Q., 2018a. Végétal, in Bourg D., Papaux A. (ed.). *Dictionnaire de la pensée écologique*, <https://lapenseeecologique.com/végétal-écologie-philosophie-et-ethique/> (visited on January 27, 2020).

Hiernaux Q., 2018b. Biologie et mésologie : une perspective végétale, in Llored J.-P., Augendre M., Nussaume Y. (ed.). *Actes du colloque de Cerisy. La mésologie, un autre paradigme pour l'anthropocène*, Paris, Hermann, 299-306.

Hiernaux Q., 2019. History and epistemology of plant behavior: a pluralistic view?, *Synthese, special issue on the biology of behaviour: explanatory pluralism across the life sciences*, 198(4), 3625-3650.

Hiernaux Q., 2021. Differentiating Behaviour, Cognition, and Consciousness in Plants. *Journal of Consciousness Studies*, 28 (1-2), 106-135.

Hodge A., 2009. Root decisions. *Plant, Cell and Environment*, 32(6), 628-640.

Holopainen J.K., Blande J.D., 2012. Molecular Plant Volatile Communication, in López-Larrea C. (ed.). *Sensing in Nature*. New York, Austin, Springer Science and Business Media: Landes Bioscience.

Jacob F., 1981. *Le jeu des possibles*, Paris, Fayard.

Jonas, H., 1966. *The Phenomenon of Life: Toward a Philosophical Biology*, New York, Harper and Row.

Karban R., Yang L.H., Edwards K.F., 2014. Volatile Communication between Plants that Affects Herbivory: A Meta-Analysis. *Ecology Letters*, 17(1), 44-52.

Karban, R., 2008. Plant behaviour and communication. *Ecology Letters*, 11, 727-739.

Kawecki T.J., 2010. Evolutionary ecology of learning: insights from fruit flies. *Population Ecology*, 52, 15-25.

Kelly C.K., 1992. Ressource Choice in *Cuscuta europea*. *Proceedings of the National Academy of Sciences of the United States of America*, 89, 12194-12197.

La Mettrie J. Offray de, 2003. *L'homme-plante*, Paris, le Corridor bleu (1st ed. 1748).

Lalande A., 1996. *Vocabulaire technique et critique de la philosophie*, Paris, PUF.

Lapidge K.L., Oldroyd B.P., Spivak M., 2002. Seven suggestive quantitative trait loci influence hygienic behavior of honey bees. *Naturwissenschaften*, 89, 565-568.

Larousse P., 2002. *Le Petit Larousse illustré*, Paris, Larousse.

Le Neindre P., Dunier M., Larrère R., Prunet P. (eds), 2018. *La conscience des animaux*, Versailles, éditions Quæ.

Legg S., Hunter M., 2007. A collection of definitions of intelligence. *Frontiers in Artificial Intelligence and Applications*, 157, 17-24.

Lenne C., 2014. *Dans la peau d'une plante*, Paris, Belin.

Linné C., 2005. *Fondements botaniques, qui, comme Prodrome à de plus amples travaux livrent la théorie de la science botanique par brefs aphorismes*. Trad. fr. Dubos G. et Hoquet T., Paris, Vuibert (1st ed. 1736).

Lovejoy A., 1964. *The Great Chain of Being: A Study of the History of an Idea*. Cambridge, MA, Harvard University Press (1st ed. 1936).

Magnin-Gonze J., 2009. *Histoire de la botanique*, Paris, Delachaux et Niestlé.

Mahall B.E., Callaway R.M., 1991. Root communication among desert shrubs. *Proceedings of the National Academy of Sciences of the United States of America*, **88**, 874-876.

Mahall B.E., Callaway R.M., 1992. Root communication mechanisms and intracommunity distributions of two Mojave desert shrubs. *Ecology*, 73, 2145-2151.

Mahall B.E., Callaway R.M., 1996. Effects of Regional Origin and Genotype on Intraspecific Root Communication in the Desert Shrub *Ambrosia dumosa* (Asteraceae). *American Journal of Botany*, 83, 93-98.

Maher C., 2017. *Plant Minds: A Philosophical Defense*, New York, Routledge.

Mancuso S., Viola A., 2015. *Brilliant Green. The surprising History and Science of Plant Intelligence*. Trad. Benham J., Washington DC, Island Press.

Marder M., 2012. Plant intentionality and the phenomenological framework of plant intelligence. *Plant Signaling & Behavior*, 7(11), 1-8.

Marder M., 2013a. *Plant-Thinking a Philosophy of Vegetal Life*, New York, Columbia University Press.

Marder M., 2013b. Is it ethical to eat plants? *Parallax*, 19(1), 29-37.

Margulis L., Sagan D., 1995. *What is life?*, London, Weidenfield and Nicolson Ltd.

Markel K., 2020. Lack of evidence for associative learning in pea plants. *Elife*, 9, e57614. https://doi.org/10.7554/eLife.57614.

Michaelian K., Sutton J., 2017. Memory, in Zalta E.N. (ed.). *The Stanford Encyclopedia of Philosophy* (Summer 2017 Edition), <https://plato.stanford.edu/archives/sum2017/entries/memory/> (visited on January 27, 2020).

Miller E.P., 2002. *The Vegetative Soul: From Philosophy of Nature to Subjectivity in the Feminine*, Albany, NY, State University of New York Press.

Morgan C.L., 1894. *An Introduction to Comparative Psychology*, Londres, W. Scott.

Myers N., 2015. Conversations on Plant Sensing: Notes from the Field. *Nature Culture*, 3, 35-66.

Okano H., Hirano T., Balaban E., 2000. Learning and memory. *Proceedings of the National Academy of Sciences of the United States of America*, 97(23), 12403-12404.

Oldroyd G., 2013. Speak, friend, and enter: Signalling systems that promote beneficial symbiotic associations in plants. *Nature Reviews Microbiology*, 11, 252-263.

Pelt J.-M., 1996. *Les langages secrets de la nature : la communication chez les animaux et les plantes*, Paris, Fayard.

Pieron J., 2009. De l'analytique existentiale à la zoologie privative : le problème de la différence anthropologique et l'amorce du « tournant ». *Alter revue de phénoménologie*, 19, 195-212.

Pieron J., 2010. Monadologie et/ou constructivisme ? Heidegger, Deleuze, Uexküll. *Bulletin d'analyse phénoménologique*, 6(2), 86-117.

Pollan M., 2013. The intelligent Plant. *The New-Yorker*, December 23, <http://www.newyorker.com/reporting/2013/12/23/131223fa_fact_pollan?currentPage=all> (visited on January 27, 2020).

Pouteau S., 2014. Beyond "second animal": Making sense of plant ethics. *Journal of Agricultural and Environmental Ethics*, 27(1), 1-25.

Pouteau S., 2018. Plants as open beings: from aesthetics to plant-human ethics, in Kalhoff A., Di Paola M. et Schörgenhumer M. (eds). *Plant Ethics: Concepts and applications*, London, New York, Routledge, 82-97.

Pradeu T., 2009. *Les Limites du soi : Immunologie et identité biologique*, Montréal, Presses de l'Université de Montréal, Paris, Vrin.

Proust J., 2003. *Les animaux pensent-ils ?*, Paris, Bayard.

Raja V., Silva P.L., Holghoomi R. *et al.*, 2020. The dynamics of plant nutation. *Scientific Reports*, 10, 19465.

Raven, P. H., Evert, R., Eichorn, S., 2013. *Biology of Plants*, New York: W. H., Freeman and Company.

Rhoades D.F., 1983. Response of alder and willow to attack by tent caterpillars and webworms: evidence for pheromonal sensitivity of willows, in Hedin P.A. (ed.). *Plant Resistance to Insects*, Washington DC, American Chemical Society, 55-56.

Rothenbuhler W.C., 1964. Behavior Genetics of Nest Cleaning in Honey Bees. IV. Responses of F1 and Backcross Generations to Disease-Killed Blood. *American Zoologist*, 45, 111-123.

Rowlands M., 2010. *The new science of the mind. From extended mind to embodied phenomenology*, Cambridge, MA, MIT Press.

Sachs J. von, 1890. *History of Botany* (1530-1860) (transl. Garnsey and Balfour), Oxford, Clarendon Press.

Schulze E.-D., Mooney H.A., 1994. *Biodiversity and Ecosystem function*, Berlin, Springer Verlag.

Selosse M.-A., 2017. *Jamais seuls*, Arles, Actes Sud.

Selosse M.-A., Richard F., Courty P.-E., 2007. Plantes et champignons : l'alliance vitale. *La Recherche*, 441, 58-61.

Silvertown J., Gordon D.M., 1989. A Framework for Plant Behavior. *Annual Review of Ecology and Systematics*, 20, 349-366.

Simondon G., 2004. *Deux leçons sur l'animal et l'homme* (1963-1964), Paris, Ellipses (1st ed. 1964).

Simonetti V., Bulgheroni M., Guerra S. *et al.*, 2021. Can Plants Move Like Animals? A Three-Dimensional Stereovision Analysis of Movement in Plants. *Animals*, 11, 1854.

Singer, P., 1975. *Animal Liberation: a New Ethics for Our Treatment of Animals*, New York, Haper Collins.

Song Y.Y., Zeng R.S., Xu J.F., Li J., Shen X., Yihdego W.G., 2010. Interplant communication of tomato plants through underground common mycorrhizal networks. *PLOS ONE*, 5(10).

Staddon J., 1983. *Adaptative behaviour and Learning*, Cambridge, MA, Cambridge University Press.

Stenhouse D., 1974. *The Evolution of Intelligence: A General Theory and Some of its Implications*, London, George Allen and Unwin.

Sternberg R., Sternberg K., 2017. *Cognitive psychology*, 7th edition. Boston, MA, Cengage Learning.

Stone C., 1972. Should trees have standing? Towards legal right for natural objects. *Southern California Law Review*, 45, 450-501.

Struik P.C., Yin X., Meinke H., 2008. Plant neurobiology and green plant intelligence, science, metaphors and nonsense. *Journal of the Science of Food and Agriculture*, 88, 363-370.

Sultan S.E., 2015. *Organism and Environment*, Oxford, UK, Oxford University Press.

Taiz L., Alkon D., Draguhn A., Murphy A., Blatt M., Hawes C., Thiel G., Robinson D., 2019. Plants Neither Possess nor Require Consciousness. *Trends in Plant Science*, 24(8), 677-687.

Tassin J., 2016. *À quoi pensent les plantes ?*, Paris, Odile Jacob.

Thellier M., 2015. *Les plantes ont-elles une mémoire ?*, Versailles, éditions Quæ.

Théophraste, 2010. *Recherches sur les plantes : à l'origine de la botanique*, Paris, Belin.

Théophraste, 2012. *Les causes des phénomènes végétaux*. Paris, Les Belles Lettres.

Thompson E., 2007. *Mind in Life*. Cambridge, MA, Belknap.

Tinbergen, N., 1963. On aims and methods of ethology. *Zeitschrift für Tierpsychologie*, 20, 410-433.

Trewavas A., 2002. Mindless mastery. *Nature*, 415(6874), 841.

Trewavas A., 2003. Aspects of plant intelligence. *Annals of Botany*, 92, 1-20.

Trewavas A., 2004. Aspects of plant intelligence: an answer to Firn. *Annals of Botany*, 93, 353-357.

Trewavas A., 2005. Green Plants as Intelligent Organisms. *Trends in Plant Sciences*, 10(9), 413-419.

Trewavas A., 2007. Plant neurobiology—all metaphors have value. *Trends in Plant Science*, 12, 231-233.

Trewavas A., 2014. *Plant Behaviour and Intelligence*, Oxford, Oxford University Press.

Tulving E., 1985. How many memory systems are there? *American Psychologist*, 40, 385-398.

Turgeon R., Webb J.A., 1971. Growth inhibition by mechanical stress. *Science*, 174, 961-962.

Turkington R., Harper J.L., 1979. The growth, distribution and neighbor relationships of *Trifolium repens* in a permanent pasture. IV. Fine scale differentiation. *Journal of Ecology*, 67, 245-254.

Turlings T., Tumlinson J., Lewis J., 1990. Exploitation of herbivore-induced plant odors by host-seeking parasitic wasps. *Science*, 250(4985), 1251-1253.

Uexküll J. von, 1909. *Umwelt und Innenwelt der Tiere*, Berlin, J. Springer.

Uexküll J. von, 1984. *Mondes animaux et mondes humains*. Trad. fr. Muller P., Paris, Denoël (1st ed. 1934).

Uexküll J. von, 2010. *Milieu animal et milieu humain*. Trad. fr. Freville C.-M., Paris, Payot et Rivages (1st ed. 1934).

Uexküll, J. von, 2010. *A Foray into the Worlds of Animals and Humans with a Theory of Meaning* (transl. J. D. O'Neil), Mineapolis, University of Minnesota Press.

Van Duijn M., Keijzer F., Franken D., 2006. Principles of minimal cognition: casting cognition as sensorimotor coordination. *International Society for Adaptive Behavior*, 14(2), 157-170.

Wilkins M.B., 1995. Are plants intelligent?, in Day P., Catlow C. (eds). *Bicycling to Utopia*, Oxford, UK, Oxford University Press.

Witzany G., 2008. The biosemiotics of plant communication. *The American Journal of Semiotics*, 24(1-3), 39-56.

Xiong T.-C., Bourque S., Mazars C., Pugin A., Ranjeva R., 2006. Signalisation calcique cytosolique et nucléaire et réponses des plantes aux stimulus biotiques et abiotiques. *Medecine/Sciences*, 22(12), 1025-1028.

Zebelo S., Maffei M., 2012. Signal transduction in plant-insect interactions: From membrane potential variations to metabolomics, in Volkov A. (ed.). *Plant Electrophysiology: Signaling and Responses*, Berlin, Heidelberg, Springer, 143-172.

In the same collection

**Gouverner la biodiversité
ou comment réussir à échouer**
V. Devictor, L. Jacquet, 2023, 82 p.

**Science et émotion. Le rôle de l'émotion
dans la pratique de la recherche**
E. Petit, 2022, 80 p.

Des choses de la nature et de leurs droits
S. Vanuxem, 2020, 116 p.

**Les harmonies de la Nature à l'épreuve
de la biologie. Évolution et biodiversité**
P.-H. Gouyon, 2020, 86 p.

Climatiser le monde
S.C. Aykut, 2020, 82 p.

La permaculture ou l'art de réhabiter
L. Centemeri, 2019, 152 p.

**De la protection de la nature au pilotage
de la biodiversité**
P. Blandin, 2019, 182 p.

Gouverner un monde toxique
S. Boudia, N. Jas, 2019, 124 p.

Print for you in Germany by Books on Demand.

Legal deposit: July 2023